Exploring STATISTICS
WITH THE TI-81

GAIL BURRILL
Whitnall High School
Greenfield, WI

PATRICK HOPFENSPERGER
Homestead High School
Mequon, WI

Edited by Demana and Waits

ADDISON-WESLEY PUBLISHING COMPANY

Reading, Massachusetts • Menlo Park, California • New York
Don Mills, Ontario • Wokingham, England • Amsterdam • Bonn
Sydney • Singapore • Tokyo • Madrid • San Juan • Milan • Paris

Many of the designations used by manufacturers and sellers to distinguish their products are claimed as trademarks. Where those designations appear in this book, and Addison-Wesley was aware of a trademark claim, the designations have been printed in initial caps or all caps.

Copyright ©1993 by Addison-Wesley Publishing Company, Inc.

All rights reserved. No part of this publication may be reproduced, stored in a retrieval system, or transmitted, in any form or by any means, electronic, mechanical, photocopying, recording, or otherwise, without the prior written permission of the publisher. Printed in the United States of America.

ISBN 0-201-52432-5

2 3 4 5 6 7 8 9 10 MU 96959493

Acknowledgments

We gratefully acknowledge the work of James Schultz, who served as consultant for these materials on teaching statistics. His advice and suggestions were invaluable. We are also grateful to Bert Waits and Frank Demana for their leadership in originating the project and to the students at Whitnall High School in Greenfield, Wisconsin, and Homestead High School in Mequon, Wisconsin, for their help and support in testing the materials in our statistics classes.

Contents

Introduction   vii

Chapter 1   Univariate Data   1
    1.1   The Shape of Things   1
    1.2   All About Average   7
    1.3   Variation and the Median   11
    1.4   Variation and the Mean   14
    1.5   Making Decisions   17

Chapter 2   Bivariate Data   23
    2.1   Scatterplots   23
    2.2   Plots over Time   27
    2.3   The Line $y = x$ 30
    2.4   Fitting a Line to Data   34

Chapter 3   Curve Fitting   39
    3.1   Correlation   39
    3.2   Variation and Linear Data   43
    3.3   Residuals   47
    3.4   Nonlinear Models   49
    3.5   Curve Fitting (Optional)   56

Chapter 4   Simulation   61
    4.1   Simulation — Probability Equal to 1/2   61
    4.2   Probability Not Equal to 1/2   66
    4.3   Unknown Number of Events   70

Chapter 5   Probability Distributions   73
    5.1   Methods of Counting   73
    5.2   Binomial Probability Distribution   76

　　　　5.3　Discrete Probability Distributions　　80
　　　　5.4　Normal Distribution　　86
　　　　5.5　Area Under the Normal Curve　　89

Chapter 6　Inference　97
　　　　6.1　Sampling Distribution　　97
　　　　6.2　Sampling Distribution of Sample Means　　102
　　　　6.3　Testing a Claim About a Mean　　105
　　　　6.4　Chi-Square　　110
　　　　6.5　Interpreting Chi-Square　　114

References　118

Introduction

The TI-81 can be an effective tool when doing statistics. This book shows how the TI-81 can be used to analyze data involving one or two variables, to graph data, to fit curves to data, to simulate probability experiments, to investigate probability distributions, and to make statistical inferences.

Some general information will help make this tool easy to use. The home screen for statistics is reached by pressing [2nd] [STAT]. This gives three options, which are accessed by using the arrow keys: CALC (Calculate) mode, DRAW mode, or DATA mode. The CALC mode calculates the standard statistics for one-variable data and regression lines for bivariate data. The [VARS] key will access any of these calculations for use outside the statistics menu. The calculations, however, must first be done in the statistics mode. The DRAW mode gives three choices: a histogram for one variable, a scatterplot for two variables, and a line that connects the points by segments in the order in which they were entered. Data is entered, edited, and cleared using the DATA mode. All data remains in memory until data is cleared as explained below. To enter data while in a program, [2nd] [{x}] and [2nd] [{y}] can be used. (See page 7–16 of the TI-81 Guidebook.)

Two other menus accessed using the [MATH] key are useful for statistics and probability. The NUM (Number) menu can be used to round numbers or apply the greatest integer function. The PRB (Probability) menu can be used to generate random numbers and to find combinations and permutations. [2nd] [Quit] can be used to exit from any routine. Other functions of the calculator such as the trace and zoom features can also be used with statistical data.

Mistakes can easily be corrected by using the [DEL] (delete) and [INS] (insert) keys. A common syntax error is to enter two commands such as <1-var> and <Hist> without performing either. Use an arrow key to return the cursor to the beginning of a command and press [DEL] to eliminate that command. Then move the cursor to the right of the next command using the right arrow key, or delete that command also.

There are three ways to clear information from the calculator. To clear the home screen, press [CLEAR]. To clear statistical data, select <ClrStat> from the STAT DATA menu. To clear the graph of a function, press [2nd] [DRAW] <ClrDraw>. Make sure all unwanted functions are deselected from the [y=] menu before using STAT DRAW.

Enjoy using your calculator with statistics. The rewards can make it worthwhile.

1

Univariate Data

1.1 The Shape of Things

Statistics can be thought of as making sense out of data. One of the first steps in making sense from data is to create a graph or picture of the data. The April 1990 *Consumer Reports* listed the fuel economy in miles per gallon for the following small cars.

Car	MPG	Car	MPG
Eagle Summit	33	Pontiac Le Mans	28
Ford Festiva	37	Subaru Justy	34
Ford Escort	33	Subaru Loyale	25
Honda Civic	32	Toyota Corolla	29
Mazda 323 Protegé	32	Toyota Tercel	35
Mercury Tracer	26	Volkswagen Jetta	26
Nissan Sentra	33		

Copyright 1990 by Consumers Union of United States, Inc., Mount Vernon, NY 10553. Excerpted by permission from *Consumer Reports,* April, 1990.

One way to display the data is to use a histogram, which divides the values into equal intervals and displays the count or frequency of the data points in each interval. To create a histogram of the miles per gallon estimates, the first step is to clear any previous data. Press [2nd] [STAT] and use [◄] to select DATA. Use [▼] to select <ClrStat>, and press [ENTER] twice. The screen will read "Done," which means any previously entered data has been cleared. To enter the new data, press [2nd] [STAT] and select DATA. Move the cursor to Edit to enter the data. Press [ENTER].

To begin, enter each data point as it appears on the list with the miles per gallon as the *x* value and the frequency of that data point as the *y* value. To enter the miles per gallon for the Eagle Summit, press 33 [ENTER] [ENTER] ; the cursor will be at

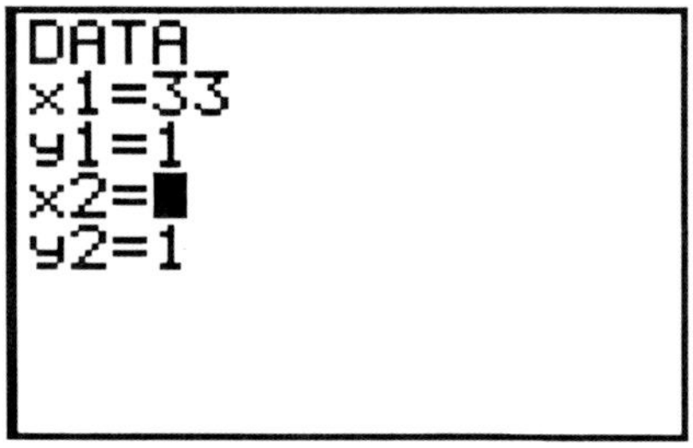

2　Chapter 1　Univariate Data

```
DATA
x1=33
y1=1
x2=■
y2=1
```

FIGURE 1.1

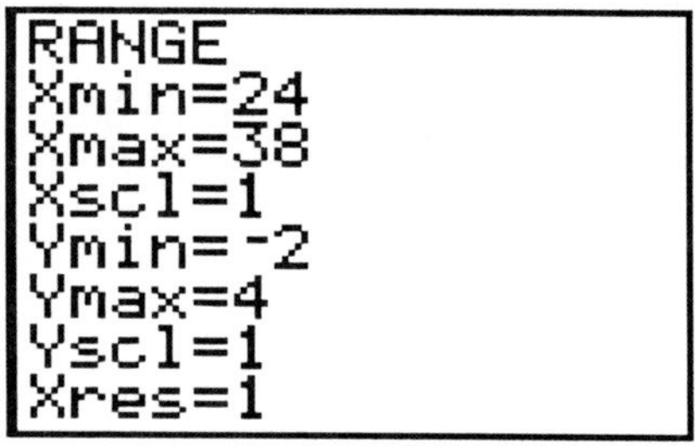

```
RANGE
Xmin=24
Xmax=38
Xscl=1
Ymin=-2
Ymax=4
Yscl=1
Xres=1
```

FIGURE 1.2

x_2. See Fig. 1.1. Continue to enter all of the data points. To make sure you have entered the right numbers, it is a good idea to read your entries on the calculator before proceeding. If you make a mistake, use the left arrow key to move the cursor to that point, and press the delete key. If you would like to omit an entire data point, move the cursor to the equals sign and use the delete key.

The next step is to find the proper viewing rectangle. Determine the range of the data (25 to 37) and estimate the frequency (from 0 to 4). To draw the histogram, press $\boxed{\text{RANGE}}$ and use the following range values: the minimum x as 24, the maximum x as 38, the x scale as 1, the minimum y as -2 (be sure to use the negative sign $\boxed{(-)}$ and not the minus sign $\boxed{-}$), the maximum y as 4, and the y scale as 1. See Fig. 1.2. (In the rest of the text, this will be written in the form [24, 38] by [-2, 4], xscl = 1, yscl = 1.)

Once the data has been entered and the appropriate range set, you are ready to draw the histogram. Press $\boxed{\text{2nd}}$ [STAT]; select DRAW and <Hist> from the menu. Press $\boxed{\text{ENTER}}$. "Hist" will appear on the screen. Press $\boxed{\text{ENTER}}$ again, and the histogram appears.

Now press any arrow key. A blinking cursor will appear in the middle of the histogram, with the coordinates of that point given at the bottom of the screen (Fig. 1.3). To find the height of a bar, use the arrow keys to move the cursor to the center of the top of a bar. The x-coordinate will give the miles per gallon, and the y-coordinate will give the frequency. Alternatively the frequencies can be determined by comparing the height of the bar to the scale of tick marks on the left of the screen. Note that by choosing minimum y to be -2, space is provided to print the x- and y-coordinates.

The cursor on the bar in Fig. 1.4 indicates one car had 29 miles per gallon. To sketch the histogram on your paper, determine the labels on the axes. Use the cursor and the range values, and make your scale accordingly. Remember that the data should be in terms of whole numbers, so round the decimals to the appropriate whole number.

The height of a bar represents the frequency. The area of the bar is the fraction this set of data points is to the total. The group of three cars with gas mileage less than 27 miles per gallon seems to be approximately 1/4 of the total area. More than half of

1.1 The Shape of Things   3

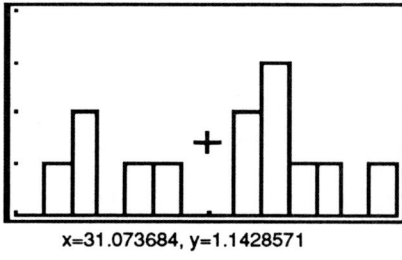

x=31.073684, y=1.1428571

FIGURE 1.3

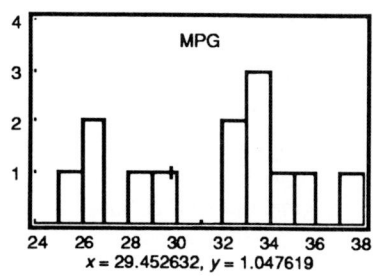

$x = 29.452632, y = 1.047619$

FIGURE 1.4

the area is from 28 to 35 miles per gallon, which means that more than half of the cars averaged from 28 to 35 miles per gallon.

An alternate method of entering the data is to organize the data first. A stem and leaf diagram of the miles per gallon can be used to order the data.

$$\begin{array}{c|l} 2 & \\ \cdot & 5\ 6\ 6\ 8\ 9 \\ 3 & 2\ 2\ 3\ 3\ 3\ 4 \\ \cdot & 5\ 7 \\ 4 & \end{array}$$

Key: 3|2 = 32 mpg

Clear the previous data by selecting DATA <ClrStat>. Press ENTER twice and the screen will read "Done." Notice that 26 occurs twice as a data point, and remember that y is the frequency, or number of times a data point occurs. Select DATA <Edit> from the [STAT] screen. Press 25 ENTER 1 ENTER 26 ENTER 2 ENTER 28 ENTER 1 ENTER , and so on. Enter all of the data using the y as the frequency.

To draw a new histogram, you must first clear the previous graph. Press 2nd [DRAW]; select <ClrDraw> ENTER ENTER . Now draw the histogram for the data you just entered. How does it compare with the histogram you drew the first time? (It is important to note this method does not give you a count of the number of data points by looking at the x subscript. In this example, the last data value (37) is x_9, but there are 13 data points. This is because there were two 26s, two 32s, and three 33s.)

Exercises

1. Sometimes different scales reveal different characteristics of the data. Investigate the graphs obtained by using different scales on the same data set. Use the data from the miles per gallon and x scales of 2, 5, and .5, and draw three different histograms. How do these histograms compare with each other and with the original one? What happens to the graph if the x minimum is changed to 20 and the x maximum to 40?

2. The prices of CD players tested by *Consumer Reports* for the March 1990 issue are given below.

Model	Price	Model	Price
Technics SL P555	$310	Radio Shack CD 1600	$250
Kenwood DP 7010	373	Sharp DX R820	245
NEC CD 730	385	Nakamichi OMS 2A	425
Onkyo DX 3500	368	Hitachi DA C70	354
Denon DCD 820	363	Magnavox CDB586	250
Magnavox CDB 630	316	Pioneer PD M610	376
Sansui CD X311	400	Sony CDP C900	390
Yamaha CDX 720	367	Denon DCM 555II	486
Pioneer PD6300	350	Sony CDP C700	344
JVC XL Z411BK	263	Onkyo DX C300	390
Sony CDP 770	306	JVC XL M401BK	291
Aiwa XC 004U	290	Technics SL PC20	254
Teac PD 480	230	Fisher DAC 198	272
Fisher AD 734	168	Toshiba XR 9028	195

Copyright 1990 by Consumers Union of United States, Inc., Mount Vernon, NY 10553. Excerpted by permission from *Consumer Reports,* March, 1990.

a) Select a scale, and draw a histogram of the prices. Make a sketch on your paper.

b) Use the histogram to determine about how much you would expect to pay if you were to buy a CD player. Explain.

c) Estimate what part of the area represents prices below $250. Approximately what percent is this? What part of the area represents prices greater than $350?

3. The average monthly pay in 1988 is given in the following table for states in the South and in the West.

South	Average monthly pay (in dollars)	West	Average monthly pay (in dollars)
Virginia	1754	Washington	1733
North Carolina	1553	Montana	1413
South Carolina	1500	Wyoming	1591
Georgia	1708	Colorado	1789
Florida	1626	New Mexico	1521
Alabama	1583	Utah	1575
Mississippi	1376	Nevada	1713
Tennessee	1600	Arizona	1698
Kentucky	1545	California	2010
Arkansas	1418	Oregon	1636
Louisiana	1610	Idaho	1470
Oklahoma	1591	Texas	1760

Source: *Statistical Abstract of the United States,* 1989

a) Draw a histogram for the average monthly wage in the South using [1300, 2100] by [-2, 6], xscl = 50, yscl = 1. Sketch it on your paper.

b) Describe the distribution of average monthly wages in the South. Use the histogram and trace functions to estimate the percent of the southern states with an average wage less than $1550.

c) Use the trace function and estimate the percent of the southern states with an average monthly wage between $1550 and $1650. Is this more or less than the percent greater than $1650?

d) Using the same scale, draw and sketch on your paper a histogram for the average monthly wages in the West. Use the histogram to determine whether the percent of western states with an average monthly wage from $1400 to $1600 is greater or less than the percent of states with a wage from $1700 to $1800.

e) Use the histograms to compare the wages in the two regions of the country.

4. The percent of infant deaths by mother's age is given in the following table. (To enter the x values, use the midpoint of the interval. Make your first and last interval equal in width to the other intervals; $x_1 = 13$ and $x_6 = 37$.)

Mother's age	Frequency
Under 15	1%
15–19	9%
20–24	26%
25–29	36%
30–34	22%
35 plus	7%

Source: *Statistical Abstract of the United States,* 1989

Sketch the histogram. Describe the distribution of the number of infant deaths in terms of mother's age. What might explain the shape of the distribution?

5. The following table gives the amount of money in thousands of dollars spent by leading advertisers on radio advertising.

$8937, $23,665, $18,906, $52,701 $2226, $5203, $3941, $0, $19,766, $23,456, $6824, $8627, $2873, $4874, $24,614, $15,004, $17, $2612
Source: *World Almanac,* 1989

a) Draw a histogram, and use it to describe the amount of money spent on radio advertising.

b) McDonald's is the company that spends $0 on radio advertising. What could account for this?

c) About what percent of the companies spent less than $10,000 on radio advertising?

Class Activity

Have students determine their heights in inches. As each student orally reports a height, enter the data into the calculator. Draw a histogram and use it to write a summary about the heights of the class.

1.2 All About Average

The Sunday newspaper lists the prices of used 19-foot Bayliner boats as follows: $4494, $4395, $7295, $11,200, $9750, $8400, $9500, $7995, $8600, $4900, $4850, $11,995, $9000, $10,300, $6300, $8500, $5000. How much would you typically expect to pay for a used Bayliner?

Sometimes it is necessary to summarize a set of data numerically. The number most often used is a measure for the center of the data. Enter the 17 boat prices. Then make a histogram of the prices, without organizing the data, in [4000, 12,000] by [-2, 4], xscl = 1000, yscl = 1. Note that $4494, $4395, $4900, and $4850 are included in the bar from $4000 to $5000.

Estimate the center by looking at the distribution (Fig. 1.5). What seems to best describe a typical boat price? One answer could be the $4000 to $5000 because that bar occurs the most often; the bar is the tallest bar, indicating it has the greatest frequency. The number that occurs the most often is called the **mode.** The mode is easy to find by looking at the data or at the histogram. Some data sets do not have one number used more than the others; sometimes the mode for a data set is not near the middle. Although the mode can be important to a person like a shoe salesman, other measures of center are often more useful. Another description for the typical price might be from $8000 to $9000, because that bar seems to be about in the middle of the data set. The number in the middle is called the **median.** A third possibility is about $7500, because it looks like that number might strike a "balance" between all of the other numbers. This balance point is called the **mean.**

To find the median, you must find the number that is in the middle of all the data points. This means the data must be ordered from smallest to largest. To do this on the calculator, bring up the statistics screen and select DATA. Use the ▼ to find <xsort>, and press ENTER. Press ENTER again, and the screen will read "Done." All of the x values have been sorted from smallest to largest. Press 2nd [STAT], select DATA, <Edit>, and check the values. The value of x_1 should be $4395 (Fig. 1.6).

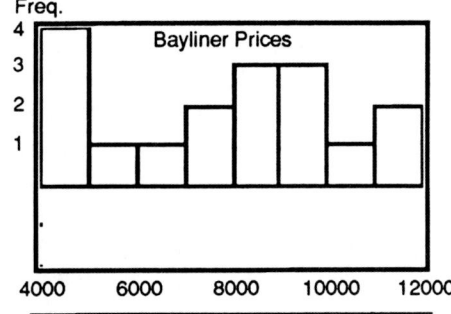

FIGURE 1.5

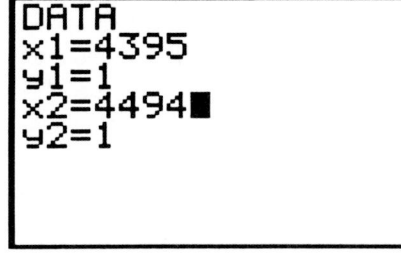

FIGURE 1.6

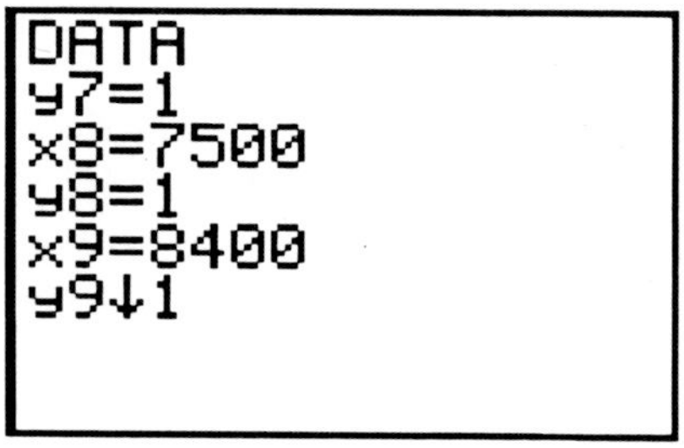

8   Chapter 1   Univariate Data

```
DATA
y7=1
x8=7500
y8=1
x9=8400
y9↓1
```

```
1-Var
x̄=7792.588235
Σx=132474
Σx²=1127364136
Sx=2437.32438
σx=2364.551967
n=17
```

FIGURE 1.7 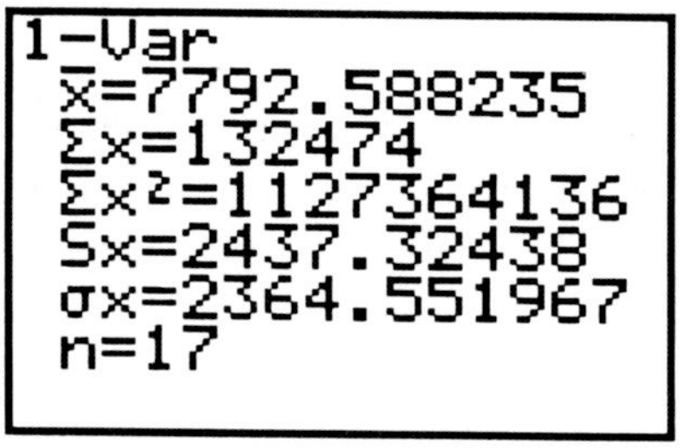   FIGURE 1.8

To find the median, recall that you entered 17 data points. The median or middle number will be the ninth ordered data point, or x_9. Use the ▼ key to find that entry (Fig. 1.7). The median is $8400. Remember this means half of the boat prices are more than $8400 and half are less than $8400.

You can also call up the median, or ninth data point, from the home screen by pressing 2nd [{x}] 9]) ENTER , and the screen will display the median. Some hints: If the number of data points is even, the median is halfway between the two middle numbers. In order to find the median, enter each data point with a frequency, or y value, of 1. (If you do not, you will have to take that into consideration when you count to the middle.)

The third measure of center is the mean, which is the sum of all the data points divided by the total number of points. The symbol for the mean is $\bar{x}$. If there are n data points this can be written in summation notation, or "sigma notation," as

$$\bar{x} = \frac{\sum_{i=1}^{n} x_i}{n}$$

The mean reflects the center in terms of the "weight" of the data. It divides the distribution into two parts that balance each other.

To find the mean, call up the STAT screen, and by default 1-var is selected from the menu. Press ENTER , and a set of statistics describing the data will appear (Fig. 1.8). The mean boat price $\bar{x}$ is $7793, which is not that far from our estimate. The line $n = 17$ at the bottom indicates the number of data points entered. (The other statistics will be explained as they are used.)

The mean is not always an appropriate measure for the center, because it is easily influenced by very large or very small data points. If the distribution is relatively symmetric (Fig. 1.9), each measure of center should describe the data fairly well. If the distribution is unbalanced or skewed (Fig. 1.10), the median will usually be more

1.2 All About Average

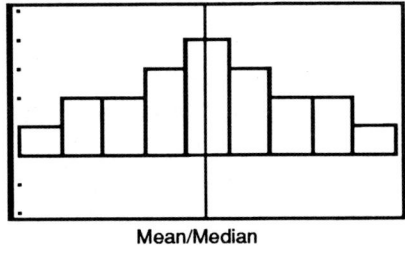

FIGURE 1.9

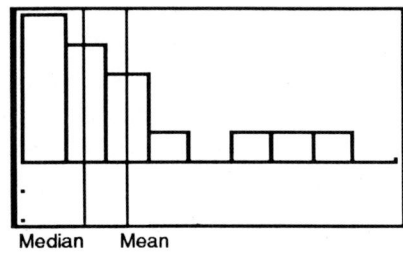

FIGURE 1.10

descriptive of the center for the data. It is important to look at the distribution before you decide which measure of center to use.

Exercises

Enter each data set into your calculator. Answer each set of questions.

1. The table contains the prices of women's walking shoes listed in *Consumer Reports* (February 1990).

Shoes	Price	Shoes	Price
Easy Spirit Mach I	$69	Rockport Prowalker 7100	$60
Saucony Instep 6200	58	Sears Winner 57402	40
FootJoy Joy Walkers 64964	59	Avia Mall Walker 310	46
Reebok Fitness Walker	60	Rockport Comfort Skimmer 3400	62
Nike Air Health Walker Plus	65	LL Bean Comfort Walkers	52
Naturalizer Performance Walker	69	Etonic Trans Am Walker K720	55
Hush Puppies Action	58		

Copyright 1990 by Consumers Union of United States, Inc., Mount Vernon, NY 10553. Excerpted by permission from *Consumer Reports*, February, 1990.

a) Draw a histogram of the prices in [40, 70] by [-2, 4], xscl = 2, yscl = 1.
b) Are most of the prices high or low? Use the histogram to estimate the mean. Is there a mode?

c) Use the histogram to estimate the percent of the prices below $58. What percent are above $68?

d) Use your calculator to find the median and mean. What does each measure represent? How do the three measures of center compare? Which seems to give the best price for a typical walking shoe?

2. The asking prices for puppies listed in the newspaper ad section are given in the following chart:

Beagle	$175	Cocker Spaniel	$220
Sheepdog	400	Cocker Spaniel	300
Bichon	149	Springer Spaniel	225
Labrador	free	Golden Retriever	275
Boxer	350	Cocker Spaniel	200
Brittany	330	Chow	100
Brittany	75		

Find the mean asking price for a puppy. Do you think this is a fair representation of the asking prices? Draw a histogram to justify your conclusion.

3. According to the World Almanac, the amount of direct foreign investment in the United States in 1988 in billions of dollars was as follows: Canada 27.3, Netherlands 48.9, Switzerland 15.8, United Kingdom 101.9, W. Germany 23.8, Japan 53.3, Middle East 5.8. Calculate the mean and median. Which is a better measure of the average amount of money foreign countries have directly invested in the United States and why?

4. The number of points scored by Portland players in an NBA playoff game is as follows: Drexler 27, Porter 18, Williams 16, Duckworth 15, Kersey 12, Petrovic 5, Robinson 4, Young 2, Cooper 1 (*Milwaukee Journal*). Explain the statement: 55% of the players scored above the average number of points per player in the game.

5. The following table lists the number of hours students in a given class work per week:

Number of hours per week	Frequency	Total
0–9	||||||||	8
10–19	|||||	5
20–29	|||||||||||||||	15
30–39	||||||	6
40–49	||	2

a) How many students were in the class?

b) Find the mean number of hours per week the students in the class worked. (Use the midpoint of the interval for the x_i.) Do you think this is a good average for your class? Why or why not?

1.3 Variation and the Median

The grand prize in a contest was a trip to a city where the average temperature was 56°. If you won the prize without knowing what city you were to visit, what clothes should you bring for the trip? Would it depend on the time of the year? San Francisco, California and Springfield, Missouri both have a mean monthly temperature of 56° and a median temperature of 57°. Would the same clothes be suitable for those cities at any time during the year? The average monthly temperatures (in Fahrenheit) are given in the following table:

Springfield, MO

Jan	Feb	Mar	Apr	May	Jun	Jul	Aug	Sep	Oct	Nov	Dec
32	36	45	56	65	73	78	77	70	58	45	36

San Francisco, CA

Jan	Feb	Mar	Apr	May	Jun	Jul	Aug	Sep	Oct	Nov	Dec
49	52	53	55	58	61	62	63	64	61	55	49

Source: *World Almanac*, 1988

It would seem helpful if we had more information about the data than just the mean or median—perhaps a number to indicate how constant the temperature of 56° is throughout the year. Enter the temperatures for Springfield in your calculator, and

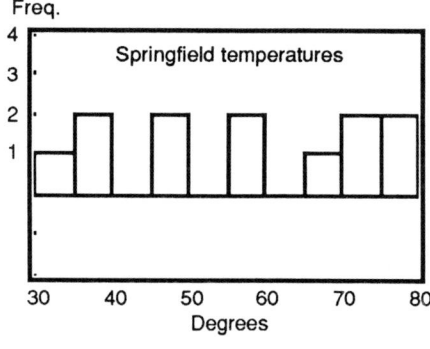

FIGURE 1.11

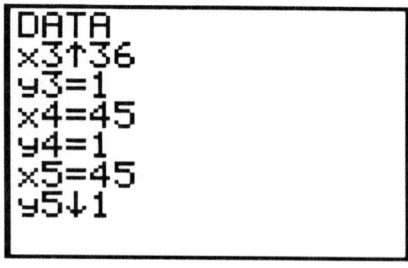

FIGURE 1.12

draw a histogram of the temperatures in [30, 80] by [-2, 4], xscl = 5, yscl = 1 (Fig. 1.11). If the measure of the center is the median, how close are the other numbers to the middle? Based on your histogram, make an estimate for an interval you think might be reasonable. One method is to find the middle half or 50% of the data. Divide the data into four equal parts or quartiles (see *Exploring Data,* Landwehr and Watkins, 1986). To find the quartiles just by counting, find the median and then find the median of each half.

To do this on your calculator, sort the x values. There are 12 data points so the median, 57, is halfway between the sixth and seventh data point. Verify this on your calculator. The lower quartile is halfway between the third and fourth data points: (36 + 45) / 2 = 40.5 (Fig. 1.12). The upper quartile is (70 + 73) / 2 = 71.5. This can be done from the home screen (2nd [{x}](3) + 2nd [{x}](4)) / 2 ENTER .

The **interquartile range** is the difference between the third and first quartile: IQR = 71.5 – 40.5 = 31. This means the middle half of the average monthly temperatures is spread over 31 degrees from 40.5° to 71.5°. Sketch the histogram. Mark the interquartile range on your histogram as illustrated in Fig. 1.13.

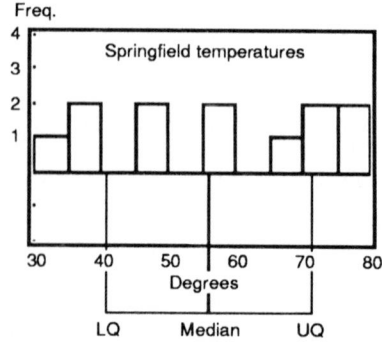

FIGURE 1.13

1.3 Variation and the Median 13

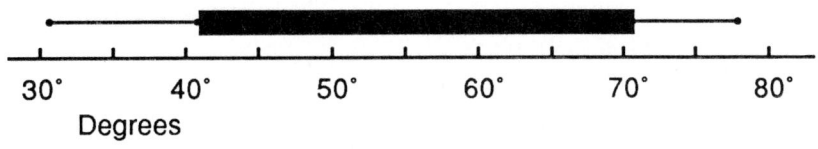

FIGURE 1.14

Another way to represent this is to draw a box plot of just the five summary numbers: the lowest extreme, the lower quartile, the median, the upper quartile and the upper extreme (Fig. 1.14). Remember you can access the quartiles by pressing [2nd] [{x}] and your data point. Compare the box plot to the histogram of the temperatures.

Exercises

1. a) Find the interquartile range for the temperatures for San Francisco. What does this tell you about the variation in the average monthly temperatures in San Francisco?

   b) Make a histogram of the temperatures for San Francisco using the same range you used for Springfield. Mark the median and interquartile range on the histogram. How does it compare with the histogram in Fig. 1.13?

   c) Draw box plots of the temperatures for San Francisco and Springfield on the same number line. Compare the temperatures in the two cities. Do the box plots reveal any more information than comparing the two histograms? Which is easier to read?

2. The prices of used Ford and Chevrolet pickups are given in the following table. Draw box plots for each set of data, and use them to compare the prices of used Ford pickups with the prices of used Chevrolet pickups. The years range from a 1962 Chevrolet for $350 to a 1988 Chevrolet for $6990, and a 1972 Ford ($1995) to a 1986 Ford ($4750).

   Ford: $1995, $650, $3500, $1950, $950, $4750, $3300, $3500, $6000, $4700, $4700
   Chevrolet: $350, $1495, $850, $2250, $6990, $9200, $2400, $2200, $1400

3. The prices of mens' walking shoes from *Consumer Reports* (February 1990) are given in the following table:

Shoe	Price	Shoe	Price
Foot Joy Joywalkers	$ 59	Nike Air Healthwalker	$ 65
Rockport Dressports Wingtip	109	Hush Puppies Mall Walker	70
		Reebok Fitness Walker	60
Avia Mall Walker 310	46	New Balance M906	110
Rockport Prowalker	70	Clarks Wallabee Air	78
Dexter Weather Walker	76	JC Penny Par Four #9539	50
Brooks StrideWalker 7100	55		

Copyright 1990 by Consumers Union of United States, Inc., Mount Vernon, NY 10553. Excerpted by permission from *Consumer Reports,* February, 1990.

a) Give an interval that would be a typical price for men's walking shoes. Justify your answer.

b) If a data point is more than 1.5 times the interquartile range above the upper quartile or below the lower quartile, it is considered an outlier (*Exploring Data,* by Landwehr and Watkins, 1986). Are there any outliers in the prices? Justify your claim.

c) Would an outlier have more effect on the mean or the median? Why?

1.4 Variation and the Mean

The mean temperature for San Francisco and Springfield was 56°. Because the mean is a measure of center that reflects a balance point for the data, the measure of variation around the mean should also reflect the weight or contribution of each data point. The **standard deviation** is a measure of the distance of each data point from the mean. The following steps can be used to calculate the standard deviation:

1. Subtract the mean from each temperature.
2. To eliminate the negative sign, square each difference.
3. Sum the squares.
4. Divide the sum by n, the number of data points. This result is called the **variance.**
5. Because the variance is not in the same scale as the original data, take the square root of the variance. The result is the standard deviation.

1.4 Variation and the Mean

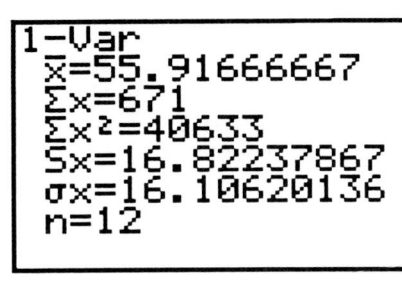

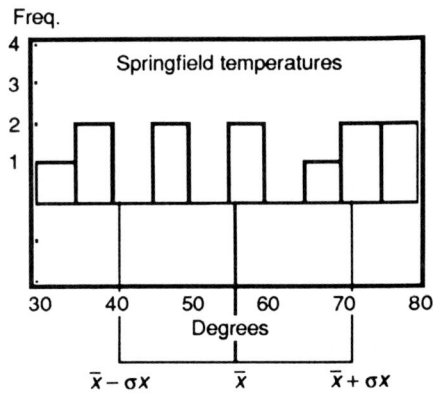

FIGURE 1.15

FIGURE 1.16

The standard deviation can also be found using your calculator. Enter the temperatures for Springfield and press [2nd][STAT], select CALC <1-var>, and press [ENTER][ENTER] (Fig. 1.15). The standard deviation σx, which is a measure of the dispersion of the temperatures around the mean temperature, is equal to 16.1 here.

One standard deviation above the mean would be 56° + 16.1° = 72.1°. One standard deviation below the mean would be 56° − 16.1° = 39.9°. Most of the temperatures are from about 40° to 72°. The standard deviation can be marked off on the histogram to get a visual concept of this interval (Fig. 1.16). (Note that if the data are originally from a sample, the divisor is $n - 1$ instead of n, and the standard deviation is given by S_x. Unless otherwise indicated, we will use σx for the standard deviation.)

Exercises

1. Use the following conditions to make a list of 12 numbers so the standard deviation is as large as possible. Check your list using your calculator.
   a) Every number must be a 1 or a 6.
   b) Every number must be a 1 or a 20.
   c) Nine numbers must be a 1 or a 20, and three numbers must be a 5.
2. Repeat Exercise 1, but make the standard deviation as small as possible.
3. Estimate whether the standard deviation of the following is closest to 0, 5, 15, or 25. Verify your estimate with your calculator.
   a) 18, 20, 28, 50
   b) 10, 15, 20, 25, 30
   c) 35, 38, 42, 50

16   Chapter 1   Univariate Data

4. The price of adult bicycle helmets was given in the May 1990, *Consumer Reports*:

Helmet	Price	Helmet	Price
Bell Quest	$51	Bell Image	$65
Performance Enduro	40	Pro-tec Mirage	41
Paramount Team Issue	58	Bell Ovation	60
Specialized	60	Troxel Adult 28-9504	40
Performance Acro	30	Spalding 2527/82533	35
Bell Spectrum	45	Brancale XP-7	47
Rhode Gear Ultralight	38	Vetta CorsaLite	49
Giro Prolight	66	Bell V-1 Pro	54
Monarch Aero-Lite	40	Giro Hammerhead	87

Copyright 1990 by Consumers Union of United States, Inc., Mount Vernon, NY 10553. Excerpted by permission from *Consumer Reports*, May 1990.

a) Draw a histogram. Mark the mean and standard deviation for the prices on the histogram. Use your histogram to estimate the percent of prices within one standard deviation from the mean.

b) Are any prices outliers? How will this affect the mean and standard deviation?

c) Write a paragraph describing the prices of bicycle helmets.

5. The amount of money spent in a week by 30 randomly selected high school seniors is given in the following table. Sketch a histogram of the results, and calculate the mean and standard deviation. Mark one standard deviation from the mean on the histogram. How well do these statistics describe the average amount of money spent by seniors in high school?

Money spent	Number of seniors
$ 5	1
10	4
15	8
17	1
20	4
25	2
30	3
40	2
50	2
70	1
90	1
100	1

6. The salaries of the employees of a small business are as follows:

$18,000, $25,000, $20,000, $20,000, $20,000, $20,000, $27,000, $18,500, $18,000, $12,000, $30,000.

a) Draw a histogram of the data, find the mean and standard deviation, and mark them on the histogram.
b) Approximately what percentage of the employees are within one standard deviation from the mean?
c) Re-enter the data as the number of thousands. For example, 18,000 should be entered as 18. Draw a histogram of the data; find the mean and standard deviation. How do the results compare with those from the original data?
d) If every employee receives a 5% raise, find the mean and standard deviation of the new salaries. How do these figures compare with the original mean and standard deviation?
e) Every employee receives a $500 bonus. Calculate their total salaries, and find the mean and standard deviation. How do these figures compare with the original mean and standard deviation?
f) Describe what seems to happen to the mean and standard deviation when the data points are each multiplied by a constant. Describe what happens when a constant is added to each data point.

1.5 Making Decisions

Often people who use statistics routinely calculate numerical measures and use the results to make conclusions. It is important, however, to look carefully at the data and the shape of the distribution and analyze what messages the data are conveying. Not all distributions have a clear center. A large standard deviation does not reveal the presence of several centers; it could just mean a large spread in the data. If numerical measures are considered with a histogram of the data, a clear picture often begins to emerge—possibly there are outliers or two different populations involved, each with its own center.

The median weekly earnings for women and men in selected occupations are given in the following table:

MEDIAN WEEKLY SALARIES (1988)

Job	Men's salaries	Women's salaries
Executive, administrative	$682	$430
Professional speciality	651	485
Technicians	510	384
Sales occupations	488	284
Administrative support (clerical)	418	305
Private household	NA	139
Protective service	424	347
Service	257	210
Mechanics and repairers	441	392
Construction trades	423	335
Precision production, craft, repair	477	284
Machine operators, assemblers	366	236
Transportation, material moving	394	286
Handlers, cleaners, laborers	287	237
Farming, forestry, fishing	234	201

Source: *Statistical Abstract of the United States,* 1989

Box plots of each set of data are useful to compare two distributions. Back to back histograms can also be used. Draw a histogram of each set using [100, 700] by [-2, 6] with xscl = 50, yscl = 1. Sketch the histograms with the bars for men going up and the bars for women going down (Fig. 1.17).

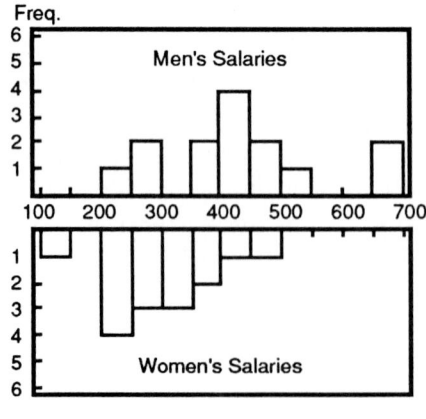

FIGURE 1.17

Exercises

1. a) What does a median weekly salary of $423 in the construction trades mean?
   b) Calculate the centers and variation for the two sets of salaries. Use these statistics and the histograms in Fig. 1.17 and write a comparison of men's and women's salaries.
2. The number of calories per one ounce serving in selected breakfast cereals is given in the following table.

Cereal	Calories	Cereal	Calories
Kellogg's All Bran	73	Kix	109
Quaker 100% Natural	121	Nabisco 100% Bran	74
Bran Flakes	88	Total Raisin Bran	89
Post Natural Fruit & Fibre	86	Wheaties	105
Honey Smacks	113	Rice Krispies	110
Quaker Puffed Rice	111	Fruit Loops	113
Product 19	112	Kellogg's Raisin Bran	89
Post Raisin Bran	84	Frosted Mini Wheats	99
Mueslix Bran	89	Life	104
Total Corn Flakes	110	Honey Nut Cheerios	102
Cheerios	106	Apple Jacks	116

Copyright 1989 by Consumers Union of United States, Inc., Mount Vernon, NY 10553. Excerpted by permission from *Consumer Reports,* October, 1989.

   a) Sketch a histogram of the calories in [70, 130] by [-2, 10], xscl = 10, yscl = 1.
   b) Change the xscl to 5, and draw the histogram again. How do the two distributions differ?
   c) The distribution seems to have two clusters. Find the intervals for the clusters, and identify the cereals in each. Do the cereals in either cluster have anything in common?
   d) Write a paragraph about the number of calories in breakfast cereals. Be sure to describe the graph, the center, and the variation.

3. The average starting salaries for college graduates in 1988 are given in the following table:

Occupation	Average starting salary
Engineering	$29,820
Accounting	24,324
Sales-marketing	22,848
Business administration	22,920
Liberal arts	22,596
Chemistry	25,692
Mathematics-statistics	26,112
Economics-finance	23,136
Computer science	27,372
Teaching	18,500

Source: *Statistical Abstract of the United States,* 1989

a) Select a career area and write a paragraph about your starting salary in terms of the total distribution.

b) Are any occupations outliers? What are some possible reasons for extremely high or low salaries?

4. The salaries for the 15 highest paid players in 1990 in the National Basketball Association and the National Baseball League are given in the following tables.

BASEBALL		BASKETBALL	
Player	Salary	Player	Salary
Robin Yount	$3,200,000	Patrick Ewing	$3,575,000
Kirby Puckett	2,700,000	Charles Barkley	2,600,000
Roger Clemens	2,600,000	A. C. Green	2,500,000
Don Mattihgly	2,500,000	Magic Johnson	2,500,000
Eddie Murray	2,492,091	Michael Jordan	2,420,000
Paul Molitor	2,433,333	Isiah Thomas	2,275,000
Kent Hrbek	2,300,000	Ralph Sampson	2,160,000
Will Clark	2,250,000	Akeem Olajuwon	2,078,737
Rickey Henderson	2,250,000	Moses Malone	2,055,000
Mike Scott	2,187,500	Alex English	2,000,000
Tom Browning	2,125,000	Chris Mullin	2,000,000
Mark Davis	2,125,000	David Robinson	2,000,000
Ted Higuera	2,125,000	Dominique Wilkins	1,975,000
Dave Winfield	2,122,890	Danny Manning	1,900,000
Eric Davis	2,100,000	Mark Jackson	1,850,000

Source: *Sports,* June 1990

a) Divide the salaries by 1000, and find the mean and standard deviation for each set. Draw back to back histograms of the two data sets.
b) Find the median and interquartile range and draw box plots of the two sets of salaries.
c) Which is a better description of the center and variation: the median and interquartile range, or the mean and the standard deviation. Why?
d) Use the plots and your statistics to decide which sport had the highest salaries for the top 15 players. Justify your answer.

5. The 1987 high school dropout rate is given for different regions of the United States in the following tables. Working in groups, compare the rates for the different regions using statistics and graphs. Rank the regions of the country in terms of dropout rates, and write a paragraph describing your conclusions. Include any possible reasons for the differences.

West	High school dropouts (%)	Central	High school dropouts (%)
Washington	21.9	North Dakota	11.6
Idaho	20.4	South Dakota	13.1
Montana	15.5	Nebraska	13.3
Oregon	29.2	Kansas	15.9
California	31.5	Minnesota	11.3
Nevada	18.7	Wisconsin	16.3
Utah	17.5	Michigan	28.6
Colorado	24.0	Iowa	13.4
Wyoming	20.4	Illinois	22.2
Arizona	30.0	Missouri	23.9
New Mexico	26.8	Indiana	24.1
Alaska	26.4	Ohio	20.8
Hawaii	15.5		

Northeast	High school dropouts (%)	South	High school dropouts (%)
Maine	21.2	Virginia	23.4
Vermont	17.3	North Carolina	30.9
New Hampshire	25.4	South Carolina	32.2
Massachusetts	24.0	Georgia	35.0
Connecticut	21.8	Florida	36.5
Pennsylvania	18.9	Kentucky	32.1
Rhode Island	28.0	Tennessee	32.8
New Jersey	20.3	Alabama	30.5
New York	33.3	Mississippi	34.4
Maryland	23.5	Arkansas	21.4
Delaware	29.0	Louisiana	38.4
Washington, DC	40.5	Oklahoma	24.2
West Virginia	22.9	Texas	34.1

Source: *Statistical Abstract of the United States,* 1989

2

Bivariate Data

2.1 Scatterplots

Sometimes you are given two sets of information such as the height and weight of the players on a football team. The graph of these data as a set of ordered pairs is called a **scatterplot**. To make a scatterplot, the first step is to decide which variable is plotted on the x-axis and which one on the y-axis. If you are trying to predict an outcome, the variable you are trying to predict is plotted on the y-axis, and the variable you are starting from on the x-axis. Sometimes it does not make any difference which variable is plotted on a given axis.

The costs for the CD players used in Chapter 1 are given in the following table along with the overall scores given by *Consumer Reports*. (A good score is a high rating.)

Model	Price	Score	Model	Price	Score
Technics SL P555	$310	97	Radio Shack CD 1600	$250	82
Kenwood DP 7010	373	96	Sharp DX R820	245	80
NEC CD 730	385	94	Nakamichi OMS-2A	425	79
Onkyo DX 3500	368	94	JVC XL Z411BK	263	90
Denon DCD 820	363	93	Sony CDP 770	306	90
Magnavox CDB 630	316	93	Aiwa XC 004U	290	89
Sansui-CD X311	400	93	Teac PD 480	230	87
Yamaha CDX 720	367	93	Fisher AD 734	168	82
Pioneer PD6300	350	92	Toshiba XR 9028	195	86

Copyright 1990 by Consumers Union of United States, Inc., Mount Vernon, NY 10553. Excerpted by permission from *Consumer Reports,* March, 1990.

24 Chapter 2 Bivariate Data

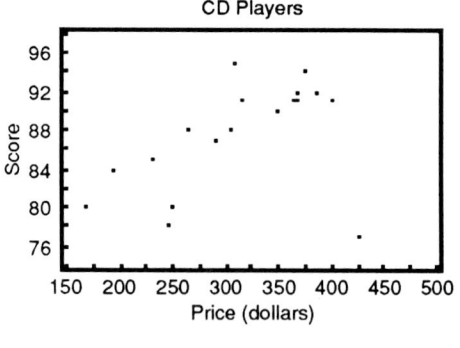

FIGURE 2.1

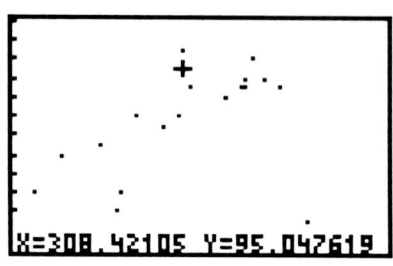

FIGURE 2.2

To look at the relation between price and score, plot the points as (price, score). To make a scatterplot of the data, press [2nd] [STAT] DATA, select <Clr-Stat> and [ENTER]. Clear any previous equations by [Y=] [CLEAR]. Press [2nd] [STAT], select DATA <Edit>, and [ENTER]. Type in the data with the price as the x_i and the ratings score as the y_i. Set the range [150, 500] by [76, 100], xscl = 25, yscl = 2. To create the graph, press [2nd] [STAT] DRAW, and select <Scatter>. Press [ENTER] twice, and the scatterplot of the data will appear (Fig. 2.1).

To find the coordinates of a single point, press any arrow. You can move the blinking cursor by pressing the arrow keys. If you put the cursor on the point, you cannot see the exact location of a point, but you can determine the approximate coordinates by lining the cursor horizontally with the point to determine the height, or y value, as in Fig. 2.2 (308.42 ≈ 310) and vertically to determine the x value (96.95 ≈ 97) as in Fig. 2.3.

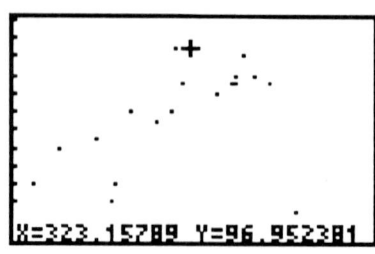

FIGURE 2.3

To find the mean test score for the CD players, you cannot use the 1-var calculations because bivariate data was entered. Instead, press [2nd] [STAT] and select CALC <LinReg>, which will internally calculate the values needed. (The statistics that appear on the screen will be explained later.) Now press [VARS], select $<\bar{x}>$, and press [ENTER] twice. The mean of the prices is $311.3, or about $311. Press [VARS], select $<\sigma x>$, and [ENTER] twice. The standard deviation is 71.6, which indicates that the typical prices for CD players tested by *Consumer Reports* were $311 ± 72 or from $239 to $383.

Exercises

1. Use the data for the CD players to answer each question.
    a) Find the mean and standard deviation for the rating scores of the CD players. What does this tell you about the typical score?
    b) Recall the scatterplot to the viewing screen. There are two points close together in the lower left. Describe the prices and rating scores of those points.
    c) Which CD player seems to be the best value for the money? Where is that player located on the graph?
    d) Were there any outliers in the data? Describe the point(s). What effect might the point(s) have on the mean price? the mean score?
    e) Comment on the sentence "If you buy an expensive CD player from this list, you will get a high quality player."

Chapter 2 Bivariate Data

2. The amount of cholesterol and total fat for snack foods is given in the chart.

Food item	Cholesterol (mg)	Total fat (gm)
Orange sherbet	14	3.8
Pudding pops, 1 pop	1	2.6
Ice milk, vanilla, soft serve	13	4.6
Ice cream, vanilla, regular	59	14.3
Ice cream vanilla, rich	88	23.7
Doughnut	20	12.0
Vanilla wafers (5 cookies)	12	3.3
Fig bars (4 cookies)	27	4.0
Chocolate brownie	14	4.0
Oatmeal cookies (4 cookies)	2	10.0
Chocolate chip cookies (4)	18	11.0
Gingerbread	1	4.0
White layer cake	3	9.0
Yellow layer cake	36	8.0
Pound cake	64	5.0
Devil's food cake	37	8.0
Corn chips	25	9.0
Chocolate pudding	15	4.0
Cream pie	8	23.0

Source: U.S. Department of Health and Human Services

Enter the data as (cholesterol, fat), and draw a scatterplot in [0, 90] by [-2, 26], xscl = 10 and yscl = 2. For (a) – (c), use the cursor to find the coordinates of the point, identify the product meeting the requirements, and indicate where the point representing the product is located on the graph. Which product is

a) highest in both cholesterol and in fat?

b) lowest in cholesterol but the highest in fat?

c) lowest in fat but the highest in cholesterol?

d) There is a small cluster of points near the origin. What is an approximate range for the amounts of cholesterol and fat of the products in this cluster? Put the cursor on the cluster, and press ZOOM , select <Zoom In>, and press ENTER twice. Redraw the scatterplot (2nd [STAT] DRAW <scatter> ENTER ENTER), and use the cursor to find the coordinates and identify the products.

2.2 Plots over Time

Plots over time are useful to look for patterns or trends. The number of people in millions below poverty level in the United States for the years 1960 to 1987 is given in the following table. Enter the data as ordered pairs (year, number). Select the viewing screen [1955, 1990] by [10, 40], xscl = 5 and yscl = 2. Press [2nd] [STAT] DRAW, select <xyLine>, and press [ENTER][ENTER]. The plot gives a picture of what is happening to the number of people below poverty level in the United States. It appears that there was a sharp drop in the number of people, but after a few years the number started to rise again. Using the cursor, you see that this rise seems to begin in 1975. The flat spot from 1986 indicates the rise seems to be slowing down (Fig. 2.4).

Year	1960	1965	1970	1975	1980	1986	1987
Number below poverty level (in millions)	39.9	33.2	25.4	25.9	29.3	32.4	32.5
Percent below poverty level	22.2	17.3	12.6	12.3	13.0	13.6	13.5

Source: *Statistical Abstract of the United States,* 1989

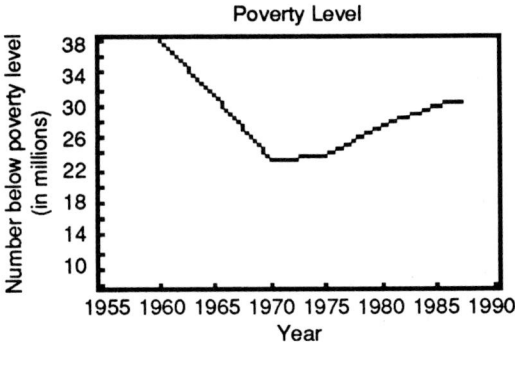

FIGURE 2.4

Exercises

1. Use the same viewing rectangle and leave the graph of the number below poverty level. Change the entries for y_i to the percent below poverty level, and draw that graph. With both graphs on the screen, answer the following:

28   Chapter 2   Bivariate Data

   a) The two lines represent the same idea, and the sharp drop in the number below the poverty level from 1965 to 1975 is apparent in both cases. Why does the corresponding increase after 1975 seem to be much smaller for the second graph?
   b) Which do you think reveals a better picture of the poverty in the United States? What is the advantage of each graph?

2. The years of life expected at birth for white and black males is given in the following chart. Make a plot over time for each on the same set of axes, and use the graphs to answer the questions.

Year	White male	Black male	Year	White male	Black male
1920	54.4	45.5	1965	67.6	61.1
1930	59.7	47.3	1970	68.0	61.3
1940	62.1	51.5	1975	69.5	63.7
1950	66.5	59.1	1980	70.7	65.3
1955	67.4	61.4	1985	71.9	67.2
1960	67.4	61.1	1988	72.1	67.4

Source: National Center for Health Statistics

   a) Describe the trend in the life expectancy at birth of a white male.
   b) Compare the life expectancy of a white male with that of a black male.
   c) Use the cursor to predict the life expectancy at birth for a white male in 1990. Of a black male in 1990.
   d) During what years was there the smallest change in life expectancy for males? What could have caused this?
   e) Was the change in life expectancy from 1920 to 1955 greater for white males or for black males? How does the graph help answer this question?

3. The times for the Daytona 500 Winners since 1972 are given in the following table:

Year	Driver	Avg. mph	Year	Driver	Avg. mph
1972	Foyt	161.5	1981	Petty	169.7
1973	Petty	157.2	1982	Allison	154.0
1974	Petty	140.9	1983	Yarborough	156.0
1975	Parsons	153.6	1984	Yarborough	151.0
1976	Pearson	152.2	1985	Elliott	172.2
1977	Yarborough	153.2	1986	Bodine	148.1
1978	Allison	159.7	1987	Elliott	176.2
1979	Petty	143.9	1988	Allison	137.5
1980	Baker	177.6	1989	Waltrip	148.5

Source: *World Almanac,* 1989

a) Make a lineplot of the times, in [70, 90] by [130, 180], xscl = 1, yscl = 5. Describe the change in the speeds for the Daytona 500 since 1972. Include the mean and standard deviation for the speeds in your discussion.

b) Use the mean ($\bar{x}$) and standard deviation (σx) from part (a) and graph the two lines $y = \bar{x} + \sigma x$ and $y = \bar{x} - \sigma x$; use $\boxed{Y=}$ and type in the constants. Plot the speeds over time on the graph. When were the speeds not within this interval for typical winning times?

c) A technique known as smoothing can be used to remove outliers or extreme points that might make it more difficult to see any pattern or trend. To determine a smoothed value for x_i, select the median of $x_{i-1}, x_i,$ and x_{i+1}; for example, to find y3, look at y2 = 157.2, y3 = 140.9, and y4 = 153.6 and select the median number 153.6. Replace y3 with 153, and continue smoothing the other data points. Use the first and last data point as listed. (See *Exploring Data* for a complete explanation.) Smooth the times for the Daytona 500, and make a line plot of the smoothed values. Describe the speeds using the smoothed plot. How does the smoothed plot differ from the original?

d) What would your prediction be for the winning speed in 1990 and 1991? How reliable are your predictions?

2.3 The Line y = x

When a scatterplot has two variables based on the same scale, it is often useful to determine if the quantities are equal or if one is greater or less than the other.

The amount of money (in thousands) spent by 13 leading advertisers on different forms of media advertising is given in the table.

Advertiser	Magazines	Newspaper	Radio	Spot radio	Network TV
Philip Morris	$243,331	$ 49,740	$ 8,937	$29,489	$330,778
Proctor & Gamble	79,611	5,576	23,665	7,906	377,552
General Motors	153,985	174,889	18,906	34,829	272,953
RJR Nabisco	105,674	20,242	2,226	8,788	209,777
PepsiCo	936	8,178	5,203	19,766	140,342
Eastman Kodak	32,884	2,866	3,941	960	145,961
Ford Motor	125,491	100,690	19,766	14,861	161,177
Anheuser Busch	11,893	11,409	23,456	43,550	186,948
Unilever	58,259	1,567	8,627	3,319	211,923
General Mills	19,983	501	2,873	1,060	133,724
Chrysler	100,446	70,166	4,874	17,345	151,569
Warner-Lambert	14,227	1,497	24,614	2,397	102,472
American Telephone & Telegraph	76,270	35,920	15,004	6,864	146,418

Source: *World Almanac*, 1989

To find out whether most of these advertisers spend more money in the two radio categories or in the newspaper category, a scatterplot can help. Clear any previously entered data and graphs, and enter the amount spent on advertising in the newspapers as x_i. Do not enter a y_i; use 1 as a default. When all of the newspaper amounts have been entered, return to the home screen by pressing [2nd] [QUIT].

To enter y_i, the amounts spent on radio, type data in the form 8937 [+] 29489 [STO] [2nd] [{y}] 1 [)] [ENTER]. (See Fig. 2.5.) This will calculate the sum and store the data point in y_i. For y_2, type 23665 [+] 7906 [STO] [2nd] [{y}] 2 [)] [ENTER]. Continue to enter the data using [2nd] [{y}] 3 for the third sum, and so on.

2.3 The Line y = x   31

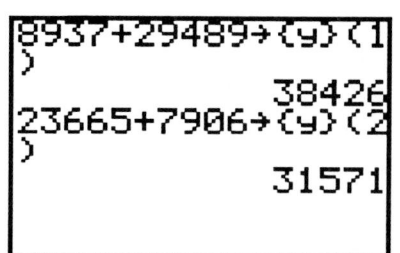

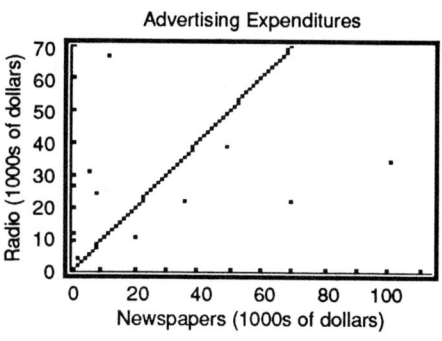

FIGURE 2.5   FIGURE 2.6

Select an appropriate range to create a scatterplot, in this example [0, 200,000] by [-10,000, 100,000], with both the xscl and yscl = 10,000. To determine whether more money is spent on radio or newspaper advertising, set the line $y = x$ by pressing [Y=][X|T]. Press [2nd] [STAT] DRAW <Scatter> [ENTER][ENTER]. Because one data point is such an outlier, it is difficult to see all of the points. Reset the range with x max as 110,000 and redraw the scatterplot (Fig. 2.6).

There are five companies below the line with an x-coordinate larger than the y; they spend more on newspaper advertising than on radio. You can trace along the line to find the x-coordinate of a particular point. General Motors, the extreme value in the upper right corner at (174,889, 53,735), also spends more on newspapers than radio advertising. There are seven companies above the line, who spend more on radio than newspaper advertising. It looks as if the top 13 companies are almost evenly divided between spending money to advertise on radio and in the newspapers.

Exercises

1. The following data give the height in inches of ninth grade students and their mothers. Enter the data as (mother's height, student's height), and draw the line $y = x$ on the scatterplot. Use the plot to answer each question.

Chapter 2 Bivariate Data

Student	Mother's height (in.)	Student's height (in.)	Student	Mother's height (in.)	Student's height (in.)
Tom	62	66	Pete	68	72
Sue	65	64	Lei	62	64
Tong	60	65	Ellie	69	65
Mark	66	70	Brian	66	72
Sally	67	66	May	64	62
Lou	65	69	Sol	70	69
Tricia	68	68	Martha	66	65
Pat	64	66	Paul	67	70
Deana	65	67	Tom	65	68
Anne	67	65	Rob	63	62

a) Explain in terms of the heights how to interpret the line $y = x$.

b) How many students are shorter than their mothers? Which part of the graph contains that information?

c) How many students are taller than their mothers?

d) Find the point representing the tallest mother and the tallest student. Where will that point lie on the graph, and to whom does it belong?

e) Find the point representing the tallest mother and shortest student. Where will that point lie, and to whom does it belong?

2. a) Using the advertising data, make a scatterplot to determine whether the companies spend more money on advertising on network TV or in print media (newspapers and magazines). Put the amount spent on TV on the horizontal axis.

b) Where is the company that spends the most on both? Identify the company.

c) Where on the plot are the companies that spend a large amount of money on TV advertising and very little on the newspapers and magazines? Identify those companies.

d) Find the mean and standard deviations for the amount the leading advertisers spend on TV and on the newspapers and magazines. Describe the typical amount spent on each form of advertising. Which form has the greater variability in spending?

3. The largest urban areas in the world are given in the following table along with their rank in 1985 and their projected rank for the year 2000. The population is in millions.

City	1985 Population	Rank	2000 Population (proj.)	Rank
Tokyo, Japan	19.0	1	21.3	3
Mexico City, Mexico	16.7	2	24.4	1
New York, U.S.	15.6	3	16.1	4
Sao Paulo, Brazil	15.5	4	23.6	2
Shanghai, China	12.1	5	14.7	7
Buenos Aires, Argentina	10.8	6	13.1	9
London, England	10.5	7	10.8	16
Calcutta, India	10.3	8	15.9	5
Rio de Janeiro, Brazil	10.1	9	13.0	10
Seoul, South Korea	10.1	9	13.0	10
Los Angeles, U.S.	10.0	11	10.9	15
Osaka, Japan	9.6	12	11.2	14
Greater Bombay, India	9.5	13	15.4	6
Beijing, China	9.3	14	11.5	13
Moscow, Soviet Union	8.9	15	10.1	17
Paris, France	8.8	16	8.8	19
Tianjin, China	8.0	17	10.0	18
Cairo, Egypt	7.9	18	11.9	12
Jakarta, Indonesia	7.8	19	13.2	8
Milan, Italy	7.5	20	8.7	20

Source: *Statistical Abstract of the United States,* 1989

a) Enter the ranks of the cities with the 1985 rank on the horizontal axis and the 2000 rank on the vertical axis. Graph the line $y = x$, and add the scatterplot of the data.

b) Where on the scatterplot are the cities whose rank is projected to increase? How many of them are there?

c) How many cities are projected to decrease in rank? Where do the points representing New York and Los Angeles lie?

d) Where will the point representing the city with the greatest change in rank lie? Which city had the greatest increase in rank? greatest decrease?

2.4 Fitting a Line to Data

Often data presented in a scatterplot seem to form a pattern that closely resembles a straight line. Just as one-variable data are sometimes summarized by using a single number such as a measure of center, the points in a scatterplot could be summarized by drawing a line through the "middle" of the points. The following table contains the number of calories and grams of fat for selected fast foods. The scatterplot of these variables for each item looks linear: the more fat, the higher the number of calories (see Fig. 2.7).

Fast food item	Grams of fat	Calories
Burger King Whopper	33	584
McDonald's Big Mac	34	572
Wendy's Big Classic	28	500
Arby's Roast Beef	19	365
Hardee's Roast Beef	17	338
Roy Rogers Roast Beef	11	335
Burger King Whaler	26	478
McDonald's Filet-O-Fish	23	415
Arby's Chicken Breast Sandwich	32	567
Burger King Chicken Tenders	12	223
Church's Fried Chicken (2 pc.)	35	487
Hardee's Chicken Filet Sandwich	20	431
Kentucky Fried Chicken (2 pc.)	31	460
Kentucky Fried Chicken Nuggets	17	281
McDonald's Chicken McNuggets	18	286
Roy Rogers Chicken (2 pc.)	35	519
Wendy's Chicken Fillet Sandwich	24	479

Copyright 1988 by Consumers Union of United States, Inc., Mount Vernon, NY 10553. Excerpted by permission from *Consumer Reports,* June, 1988.

One way to fit a line to this data is to use a **least squares regression line,** based on the means of the data points that minimizes the sum of the squares of the distance from the actual to the predicted value. To find this line using your calculator, clear the Y= screen by pressing Y= CLEAR. Clear previously drawn graphs by pressing 2nd [PROG], selecting <ClrDraw>, and pressing ENTER ENTER . Use 2nd [STAT] DATA <ClrStat> to clear the previous set of data. (Pressing 2nd [Reset] ENTER will clear all registers at once but will also clear any programs stored in memory.) Enter the ordered pairs of data with x as the number of grams of fat and y as the number of calories. Select an appropriate viewing window, in this case

2.4 Fitting a Line to Data

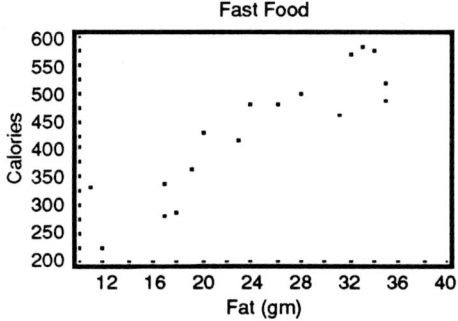

FIGURE 2.7

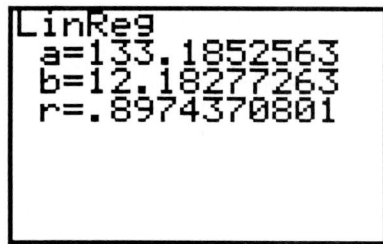

FIGURE 2.8

[10, 40] by [200, 600], xscl = 2, yscl = 25. Press 2nd [STAT] and select <LinReg>. Press ENTER twice, and the screen should look like Fig. 2.8. The standard form of the regression line uses b as slope and a as the y-intercept, so an approximate equation of a line through the data is $y = 12x + 133$. For details about the regression line, see any standard statistics text.

To graph the data and the corresponding least squares regression line, press Y = VARS and use the right arrow key to select LR. From the LR menu, use the down arrow key to select <RegEq> ENTER. This will paste the regression line onto the function y1 = . In order to see how well the points are summarized by the line, press 2nd [STAT], select DRAW, and press <Scatter>, ENTER ENTER (Fig. 2.9). This sequence of steps will plot the line and then add the scatterplot. To plot the line through the points, paste the regression line on y1 (or any y_i) as before. Move the cursor to =, and type ENTER. This locks the function y1 so it will not graph automatically on the screen. Draw the scatterplot 2nd STAT DRAW <Scatter> ENTER ENTER. Now type 2nd DRAW, and select <DrawF>. Type

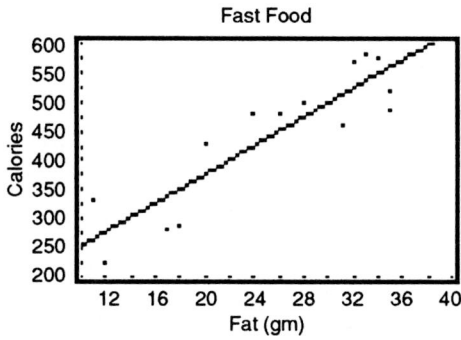

FIGURE 2.9

2nd VARS and select y1 (or y_i) ENTER ENTER, and the least squares regression line will be drawn through the scatterplot.

The trend reflected by the points seems to be fairly well summarized by the line. The least squares regression line can be used to make predictions by using the trace function. Press TRACE and the blinking cursor will appear on the line in the viewing screen. To find how many calories are in a fast food item with 28 grams of fat, use the right or left arrow to find the point on the line where the x-coordinate is 28. The corresponding y-coordinate is 474.30289, or approximately 474 calories.

Exercises

1. The median annual incomes of year round full-time workers 25 years old and over for men and women are given in the following table. Enter the data as ordered pairs with the men's salaries as the x-coordinates and women's salaries as the y-coordinates.

Year	Men	Women
1970	$ 9,521	$ 5,616
1971	10,038	5,872
1972	11,148	6,331
1973	12,088	6,791
1974	12,786	7,370
1975	13,821	8,117
1976	14,732	8,728
1977	15,726	9,257
1978	16,882	10,121
1979	18,711	11,071
1980	20,297	12,156
1981	21,689	13,259
1982	22,857	14,477
1983	23,891	15,292
1984	25,497	16,169
1985	26,365	17,124
1986	27,337	17,675
1987	28,313	18,531

Source: *Statistical Abstract of the United States,* 1989

a) Find the equation of the regression line.
b) Draw a graph of the line and the scatterplot for the ordered pairs. Make a sketch on your paper. How well does the line represent the data?

c) Write a sentence describing the slope of the line in terms of the data.
d) If the men's median salary for 1989 was $30,486, what median annual income would you expect for women?
e) Use the results of your analysis to comment on this statement taken from the *Milwaukee Journal,* February 2, 1988: "The gap (between men's and women's pay) has been steadily narrowing since 1979 when the Bureau of Labor Statistics first began collecting data on the wage differences by sex."

2. Using the annual incomes from problem 1, enter the data as ordered pairs with the years as the x-coordinates and men's salaries as the y-coordinates.
   a) Find and graph the equation of the least squares regression line.
   b) Write a sentence describing the slope of the line in terms of the data.
   c) Edit the data by replacing the y-coordinates with the women's salaries. Without clearing the previous graph, find and graph the equation of the least squares regression line.
   d) Write a sentence describing the slope of the line in terms of the data.
   e) How are the two regression lines related?

3. The average Scholastic Aptitude Test (SAT) score in mathematics in 1988 is given in the following table along with the percentage of students in that state who take the SAT test.

State	Percent of graduates taking SAT test	Math score	State	Percent of graduates taking SAT test	Math score
Alabama	8	520	Montana	20	529
Alaska	43	475	Nebraska	10	545
Arizona	22	500	Nevada	24	486
Arkansas	7	516	New Hampshire	68	487
California	44	484	New Jersey	69	469
Colorado	29	511	New Mexico	12	524
Connecticut	81	472	New York	72	469
Delaware	62	466	North Carolina	58	440
Florida	49	468	North Dakota	5	555
Georgia	63	444	Ohio	23	499
Hawaii	52	480	Oklahoma	9	522
Idaho	16	501	Oregon	50	482
Illinois	18	520	Pennsylvania	63	462
Indiana	55	458	Rhode Island	64	469

(continued)

Chapter 2 Bivariate Data

State	Percent of graduates taking SAT test	Math score	State	Percent of graduates taking SAT test	Math score
Iowa	5	577	South Carolina	57	438
Kansas	10	541	South Dakota	6	559
Kentucky	10	515	Tennessee	13	524
Louisiana	10	513	Texas	45	462
Maine	59	466	Utah	6	536
Maryland	60	475	Vermont	64	472
Massachusetts	73	474	Virginia	63	472
Michigan	13	513	Washington	37	494
Minnesota	17	531	West Virginia	14	496
Mississippi	4	519	Wisconsin	14	534
Missouri	14	519	Wyoming	12	527

Source: College Entrance Examination Board, "National SAT Scores Show Little Change for Third Straight Year, But Averages for Most Ethnic Groups Continue to Rise," 1988

a) Enter the data as ordered pairs (percent, score) for the states in the left half of the table, and calculate the regression line. Have a partner enter the data for the states from the right half, and calculate that regression line. How do your answers compare? Justify your answer.

b) Make an estimate for the regression line obtained by using all of the states.

c) If 25% of the students in a state take the SAT, what would you expect for the mean math score? If the mean mathematics score was 500, how many students would you expect to take the test?

d) What does the slope indicate about the relation between the percent of students and their SAT math score? Is the y-intercept meaningful?

e) Graph the line and the scatterplot for each column. How well do you think the line represents the data? What does this tell you about the rate of change of the scores with respect to the percent of students who take the test?

Class Activity

Collect data from students in class on the number of hours per week they study and on the number of hours per week they work. Is there any relation? Write a paragraph describing the plot.

3
Curve Fitting

3.1 Correlation

The strength of the association between two variables whose graph looks linear can be measured by an index, r, called the **correlation coefficient**. In symbols,

$$r = \frac{\Sigma(\bar{x}-x)(\bar{y}-y)}{n\sigma_x\sigma_y}.$$

It is important to recognize that r is purely a number, and it does not contain a scale factor. Because r is dependent on the mean for both x and y, it may be easily influenced by extreme values. It is also important to recognize that an association or correlation between two variables does not necessarily indicate the relation is a cause and effect relationship. There may be a strong positive correlation between the number of fouls a basketball player has and the number of points she makes, but this does not mean making fouls causes a player to score points. Both of these variables are a function of time.

The maximum value for r is +1. An r close to 1 indicates a strong positive association; as one variable increases, so does the other (Fig. 3.1a). The minimum value of r is -1. An r close to -1 indicates a strong negative association; as one variable decreases, the other increases (Fig. 3.1c). The value $r = 0$ indicates no association or relation between the variables; as values of x increase, some values of y increase and other values of y decrease (Fig. 3.1b).

Because it is possible to have a fairly large positive correlation for data that are not really linear, r alone can be misleading (Fig. 3.2). This means it is important to look at the graph of the data as well as the correlation coefficient. It is also important to observe that the graph can be misleading depending upon the scales used, so both graphical and numerical summaries are necessary to understand the data.

The quantity r^2 can be used to describe the percent of variation in y that can be predicted using the least squares regression line and a given x. If $r = .8$, $r^2 = .64$, or 64%. The least squares regression explains 64% of the observed variation in y. A correlation of .6, therefore, means the regression explains only 36% of the variation

40   Chapter 3   Curve Fitting

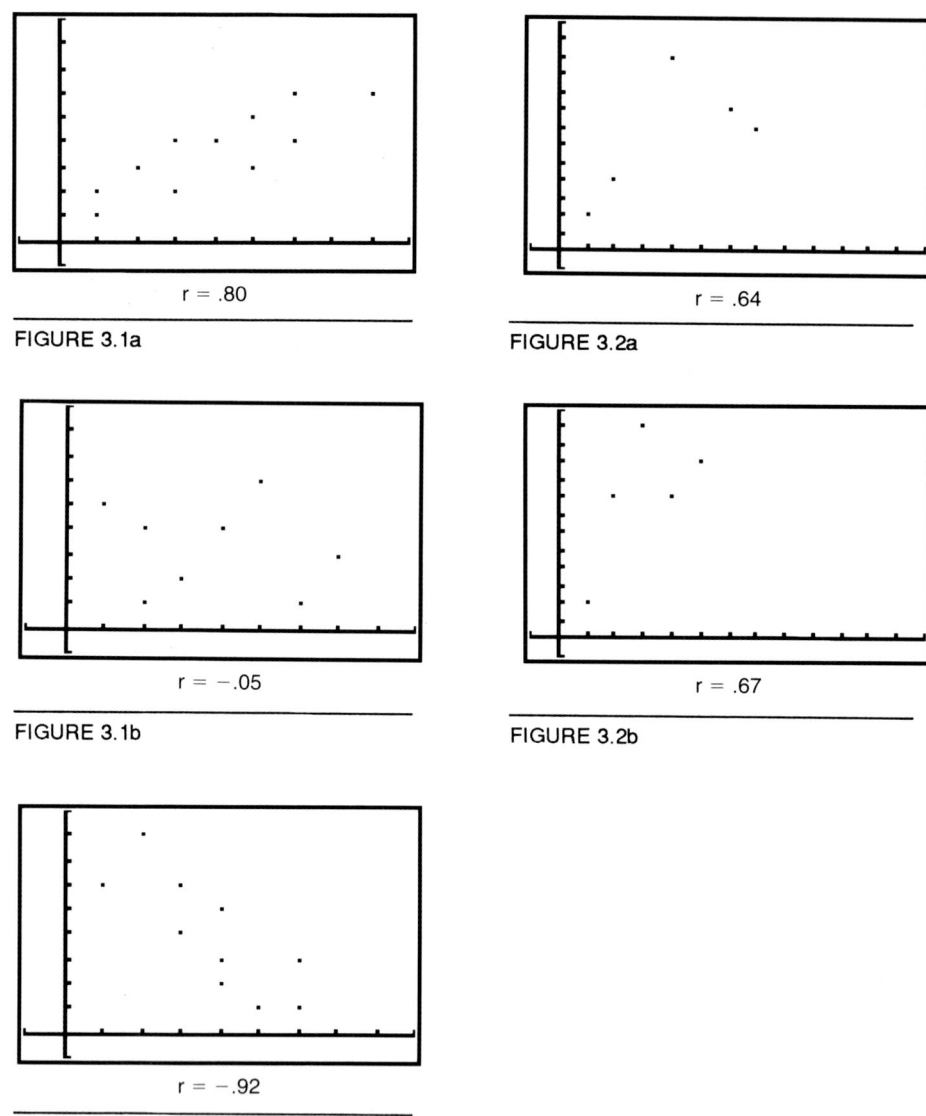

FIGURE 3.1a — r = .80

FIGURE 3.2a — r = .64

FIGURE 3.1b — r = −.05

FIGURE 3.2b — r = .67

FIGURE 3.1c — r = −.92

in y. (For a more detailed explanation see *Statistics* by Freedman, Pisani, and Purvis, or any standard statistics text.)

When <LinReg> is selected on the calculator from the 2nd [STAT] CALC menu, r is displayed on the screen along with the regression coefficients. For the fat vs. calorie data from Section 2.4, $r = .897$, or about .9, which indicates a strong positive correlation. There is a relation between grams of fat and calories in fast food items; the more fat, the more calories. The .9 indicates the least squares regression line explains 81% of the variation in the number of calories.

Exercises

1. Scientists have monitored the number of chirps per minute made by crickets and the corresponding temperature.

Number of chirps per minute:	136	165	98	110	150	210	84	158	221	178
Temperature in Fahrenheit:	72	84	68	75	80	94	60	75	92	89

   a) Make a scatterplot of the data (chirps, temperature). Find the equation of the least squares regression line and the correlation coefficient.

   b) Graph the least squares regression line and the scatterplot. What does r tell you about the relationship between the temperature and the frequency of chirps made by crickets?

   c) How well does the line seem to summarize the points?

   d) If a cricket chirps 90 times per minute, what is the predicted temperature?

   e) What percent of the variation in the change in temperature can be explained by the least squares regression line?

2. The mean number of inches of rain per year for selected cities throughout the United States is given in the following table along with the percentage of sunshine for that city.

City	Mean in. rainfall	% Sun	City	Mean in. rainfall	% Sun
Los Angeles, CA	14	73	New Orleans, LA	57	59
Salt Lake City, UT	15	70	Nashville, TN	46	57
Phoenix, AZ	7	86	Jackson, MS	49	60
Las Vegas, NV	9	84	Mobile, AL	60	67
San Francisco, CA	20	67	Atlanta, GA	61	48
Denver, CO	16	70	Charlotte, SC	66	43
Wichita, KS	31	65	Raleigh, NC	60	43
Oklahoma City, OK	31	67	Miami, FL	66	60
Albuquerque, NM	8	77	St. Louis, MO	58	36
Houston, TX	48	57	Louisville, KY	57	43
Little Rock, AR	49	63	Norfolk, VA	63	45

Source: *Greener Pastures Relocation Guide*, 1984.

a) Enter the data (rain, sun) and find the least squares regression line and the correlation coefficient. What does r tell you about the relation between the amount of sun and rain in those cities?

b) Graph the equation and the scatterplot. Does the scatterplot confirm your conclusion about the relationship between the amount of rain and sun?

c) Describe what the intercepts and the slope represent in words.

d) According to the line, if a city has an average of 25 inches of rain, what percent of the time does the sun shine?

e) According to the line, if a city has 30% sun, what is the average rainfall in that city per year?

3. The Detroit Pistons 1988–89 statistics for rebounds and total number of points scored are given in the following table.

Player	Rebounds	Total points
Aguirre	386	1511
Thomas	273	1458
Dumars	172	1186
Johnson	255	1130
Laimbeer	776	1106
Rodman	772	735
Edwards	231	555
Mahorn	496	522
Salley	335	467
Long	77	372

a) Enter the data as (rebounds, total points). Find the correlation coefficient for the relationship between the number of rebounds and total points. Explain in a sentence what this means.

b) Graph the least squares regression line and the scatterplot. How does the picture reinforce the message given by the correlation coefficient?

3.2 Variation and Linear Data

Although the correlation coefficient can be used to measure the strength of the association between two variables, the variability when a line is used to summarize data can be measured by finding an "average," or mean error. For each x in the set of data, there are two corresponding y values: the predicted y using the equation of the regression line and the actual y obtained from the data. The difference between these two values, or "error," is the vertical distance between the data point and the regression line. This difference is called a **residual.** The sum of the squares of all the errors, or residuals, divided by the number of data points is the **mean squared error.** The square root of this value can be thought of as a **standard error,** very much like finding the standard deviation for a single variable.

Consider Exercise 2 from the preceding section: the mean inches of rainfall and the percent of sun. The least squares regression line was $y = -.53x + 82.5$, where x is the mean inches of rainfall and y is the percent of sunny days. To find the standard error for this equation on your calculator, enter the following program. (It was written for program 2, but you may use any of the available program numbers.) Press ENTER at the end of each line.

Chapter 3 Curve Fitting

Enter	Display screen
[PRGM] EDIT 2 [E] [R] [R] [O] [R]	:Prgm 2:ERROR
[PRGM] I/O <Input> [ALPHA] [N]	:Input N
1 [STO] [I]	:1 → I
[PRGM] CTL <Lbl> 1	:Lbl 1
[2nd] [{x}] [ALPHA] [I] [)] [STO] [X|T]	:{x}(I) → x
[2nd] [Var] <Y1> [−] [2nd] [{y}] [ALPHA] [I] [)] [STO] [2nd] [{y}] [ALPHA] [I] [)]	:Y$_1$ − {y} (I) → {y} (I))
[PRGM] CTL <If> [ALPHA] [I] [2nd] [TEST] <=> [ALPHA] [N]	:If I = N
[PRGM] CTL <Goto> 2	:Goto 2
[ALPHA] [I] [+] 1 [STO] [I]	:I + 1 → I
[PGRM] CTL <Goto> 1	:Goto 1
[PRGM] CTL <Lbl> 2	:Lbl 2
[2nd] [STAT] CALC <LinReg>	:LinReg
[2nd] [√] [(] [VARS] Σ <y^2> [+] [ALPHA] [N] [)] [STO] [J]	:√(Σy^2/N) → J
[PRGM] I/O <Disp> [ALPHA] [J]	:Disp J
[2nd] [QUIT]	

To leave the program edit mode at any time you must press [2nd] [QUIT].

Enter the sun/rain data from Exercise 2 of the previous section, and calculate the linear regression and paste it on $Y1$. Run the program; press [PRGM], and select the program number you used to write the program. Press [ENTER] twice and ? will appear on the screen. Enter the number of data points, 22, and press [ENTER]. The standard error is approximately 7.13. For example, if a city has 40 inches of rain, the least squares regression line predicts that city has sun about 61 percent of the time with a standard error of ±7%. In other words, we can estimate it is sunny about 54% to 68% of the time.

Remember to calculate the regression before you run the program and key it into $Y1$. If you look at the data after running the program, the y values look different. To retrieve the original data, rerun the error program.

It is important to note the standard error is a mean or average and will be influenced by large differences. Points that are far from the regression line could have a large impact on the error. It is also important to recognize that unlike the correlation coefficient, a standard error must be interpreted in terms of the data. A standard error of 5 could be large for one data set and very small for another. See any statistics book for more information.

Exercises

1. The salaries and the 1988 scoring average per game for the 16 highest paid NBA basketball players are given below.

Player	Scoring avg.	Salary (in millions)
Patrick Ewing	22.7	$3.575
Charles Barkley	25.8	2.600
A. C. Green	13.3	2.500
Magic Johnson	22.5	2.500
Michael Jordan	32.5	2.420
Isiah Thomas	18.2	2.275
Ralph Sampson	6.4	2.160
Akeem Olajuwon	24.8	2.079
Moses Malone	21.7	2.055
Alex English	26.5	2.000
Chris Mullin	26.5	2.000
David Robinson	17.3	2.000
Dominique Wilkins	26.2	1.975
Danny Manning	16.7	1.900
Mark Jackson	16.9	1.850
Terry Cummings	22.9	1.846

   Source: *Sport*, June 1990

   a) Enter the data as (average, salary). Find the linear regression line and the correlation coefficient. Graph the regression line and the points. Describe the association between the scoring average and the salary of these NBA players.

   b) Use the error program to find the standard error. Remember to enter the number of ordered pairs after the ?. What does this tell you about the relation between scoring average and salary?

   c) Use the regression line and standard error to predict a salary interval for a player who averaged 25 points per game. How does the interval relate to the original data?

d) Would your results be different if Sampson were omitted? (To delete a data point, move the cursor to the equal sign for either the x or y value, and press DEL.)

2. The average per capita 1989 incomes for states in the middle east and Southeast are given in the following table along with the average price in those states for single family homes.

State	Income	Home price
Alabama	$13,679	$ 64,070
Arkansas	12,984	52,821
Delaware	19,116	92,066
Florida	17,694	76,346
Georgia	16,188	76,432
Kentucky	13,777	63,481
Louisiana	13,041	57,287
Maryland	21,020	109,191
Mississippi	11,835	58,006
New Jersey	23,764	150,302
New York	20,540	129,461
North Carolina	15,221	75,313
Pennsylvania	17,422	88,708
South Carolina	13,616	71,453
Tennessee	14,765	63,777
Virginia	18,970	100,012
West Virginia	12,529	56,149

Source: Census Bureau; Century 21 Real Estate

a) Enter the data as (income, home price). Find the least squares regression line and the correlation coefficient. Graph the regression line. How well does the line seem to summarize the data?

b) Find the standard error and describe what this tells you.

c) Use your results to summarize the relation between income and the average cost of a single family home.

d) In Wisconsin, the average income is $16,759 per year. Based on your analysis, what would you expect to pay for a single family home? Answer using an interval.

e) The actual average cost of a home in Wisconsin was given as $62,918. How does this fit with your estimate? What might explain any difference?

3. Explain why the original data are restored when the error program is rerun.

3.3 Residuals

When the program to find the standard error is used, the *y* values of the data points are transformed to the set of **residuals,** or differences between the original *y* value and the *y* value predicted by the least squares regression line (Fig. 3.3). A graph of the differences will often reveal a pattern or characteristics of the errors. If the data do not deviate from the regression model in any systematic way, the residuals usually form an irregular horizontal pattern, which ideally is symmetric about the *x*-axis.

To graph the residuals for the sun/rain data, enter the data and run the standard error program. In order to minimize rounding errors when you are calculating the residuals, use at least four significant figures when entering the regression coefficients. After running the error program, inspect the *y* values in the data using the edit command; then change the range of the *y* values to reflect the range of the residuals. For this data, use [0, 100] by [-20, 20], xscl = 10, yscl = 2.

Press $\boxed{Y=}$ and deselect or lock out any equations from the graph, then press $\boxed{2nd}$ [STAT] DRAW <Scatter> $\boxed{ENTER}$ $\boxed{ENTER}$. Figure 3.4 is the graph of the residuals. Notice they are scattered with no apparent pattern but are closer to zero for small values of *x*. This means the error in predicting from the line will increase as *x* gets larger. The graph of the residuals is important, because the standard error is a function of the mean, and it is possible for a standard error to camouflage inconsistent and extreme residuals. To restore the original data, rerun the error program.

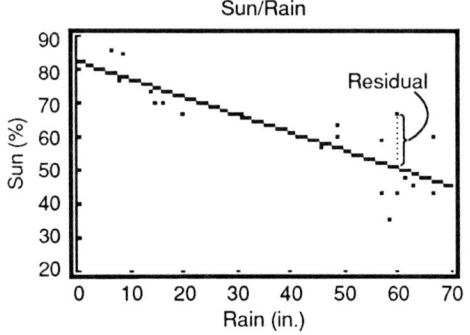

FIGURE 3.3

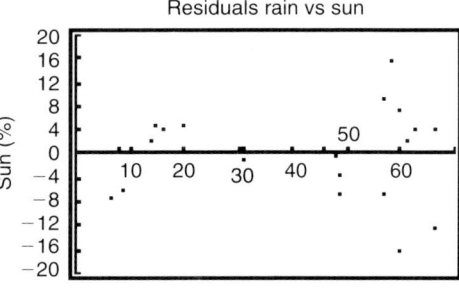

FIGURE 3.4

Exercises

1. a) Find and graph the residuals for the sun/rain data. There are three data points (two negative and one positive) that seem to have very large residuals. Use the trace to locate the coordinates of these points. What cities do these points represent? Where were these points in relation to the original regression line? Remember, the standard error was 7.13.

    b) Graph the lines $y = 7$ and $y = -7$, and inspect the residuals outside of that interval. Press [Y =] 7 [ENTER] [(-)] 7 [ENTER] [2nd] [STAT] DRAW <Scatter> [ENTER] to do this. How many data points are outside of the interval? What does this indicate about the data points that generated those residuals?

2. The ACT scores for English and Social Studies are given in the following table:

Year	English	Social Studies
1970	18.5	19.7
1975	17.7	17.4
1980	17.9	17.2
1981	17.8	17.2
1982	17.9	17.3
1984	18.1	17.3
1985	18.1	17.4
1986	18.5	17.6
1987	18.4	17.5
1988	18.5	17.4

    Source: *World Almanac,* 1990

    a) Enter the data as (English, Social Studies). Find the regression line and the correlation coefficient. What does this tell you about the data? How does the graph of the data reinforce your conclusion?

    b) Use the program on your calculator to find the standard error for the relation between English and Social Studies ACT scores. Be sure to change the regression equation before you run the program.

    c) Change the range for the y values to $[-2, 2]$, yscl = .25, and graph the residuals. Describe the plot.

    d) Find the regression line, the correlation, and the error and plot the residuals without using the data for 1970. How do your results differ from your earlier results?

3. A class of students measured their wrists and necks in centimeters. The data are given in the following table:

Wrist	Neck	Wrist	Neck
15	30	17.5	33.5
15.5	31	15	33
14	34.5	16	35
15	34	14	32
15.5	31	17	40
18	35	16.5	39.5
15	33	17.5	37.5
15.5	31	15	33
14	34.5	18	42
19	40	21	42.5
14.5	35.5	20	39
16	36.5	14.5	36.5
18	38.5		

a) Enter the data as (wrist size, neck size). Plot the data, and make an estimate of the correlation. Find the least squares regression line and the correlation coefficient. What does this information tell you about the data?

b) Find the standard error for the relation between wrist size and neck size. What does this tell you about your predictions about neck size from wrist size?

c) Graph the residuals and describe the plot. Does there seem to be any pattern?

d) Using all of your information, discuss the relation between wrist size and neck size. Include in your discussion the slope of the least squares regression line and what it means, and an interpretation of the correlation coefficient and of the standard error to determine how well the line summarizes the data.

3.4 Nonlinear Models

The graph of a set of data may have a pattern that does not look linear. The prices of used Ford Mustangs taken from the want ads of a newspaper are given in the following table along with the age of the car.

Age (years)	Price	Age (years)	Price
4	$6500	9	$2250
5	6000	11	2000
5	7000	11	3250
6	4750	12	1000
6	5000	12	1100
6	5500	13	1500
7	7000	15	750
8	4000	16	1750
8	3000	18	750
9	1500		

The equation for the linear regression is $y = -475x + 7922$ with a correlation coefficient of .86. The slope of -475 would mean a car depreciates at a rate of $475 per year. Depreciation, however, is not a constant, and it seems reasonable to assume the data are not linear. Even though the correlation is close to -1, the plot looks as if a curve would make a better model than a straight line (Fig. 3.5).

When the residuals are graphed, they are negative for very new and very old cars and positive for all but one of the rest (Fig. 3.6). There seems to be a pattern between the data points and the regression line. The standard error is $1124. Most of the differences are from -600 to 1200, and nine are outside of the interval from ±1000. A least squares linear regression does not seem to be the best model. Press 2nd [STAT] and look at the CALC menu. Three other possible models can be used for a set of data: LnReg, ExpReg, and PwrReg. These are obtained by transforming the data to "straighten" it, and then fitting a least squares line to the transformed data. (See Section 3.5.) The logarithmic transformation, LnReg, is obtained by taking the natural logarithm of x and gives equations of the form $y = a + b \ln x$. ExpReg gives equations in exponential form, $y = ab^x$, and PwrReg yields polynomial equations of the form $y = ax^b$.

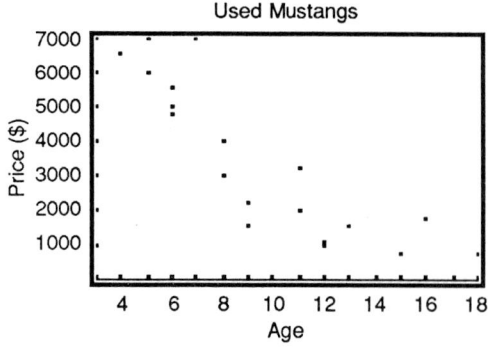

FIGURE 3.5

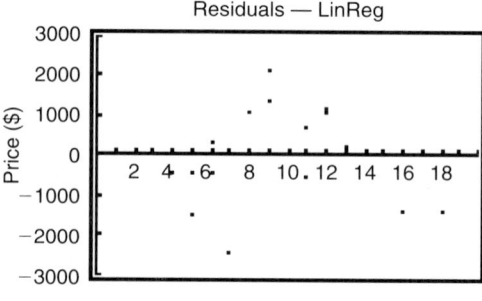

FIGURE 3.6

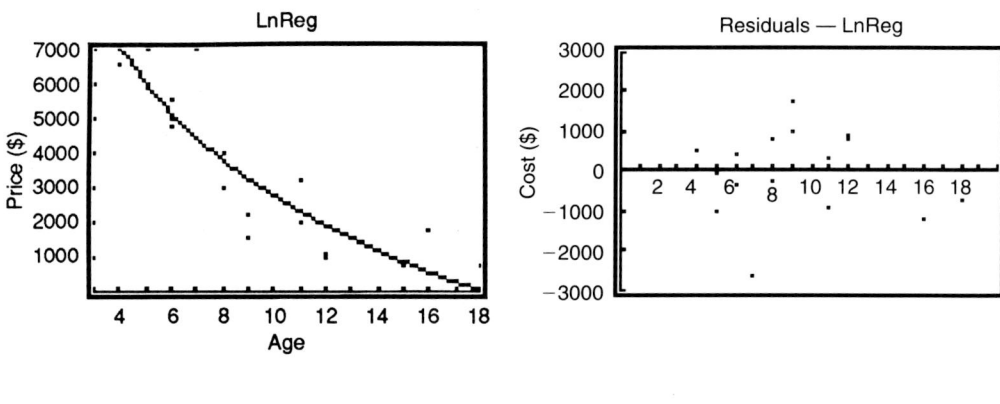

FIGURE 3.7   FIGURE 3.8

To explore logarithmic regression using the car data, return to the original data. Press [2nd] [STAT] <LnReg> [ENTER]. The quantities a and b determine the equation $y = 13{,}457 - 4640 \ln x$, and r, -.89, is the correlation coefficient for the transformed data. To determine the logarithmic equation, select <LnReg>; then press [Y =] [VARS] LR <RegEq>. To graph the data and the curve, lock the equation (move the cursor to the = by Y1 and press ENTER, press [2nd] [STAT] Draw <Scatter> ENTER [2nd] [DRAW] <DrawF> [2nd] [Y-VARS] ENTER ENTER (Fig. 3.7).

Run the error program to find the standard error, and change the range to graph the residuals. The standard error is 956.7, about $957, and the residuals do not seem to have any pattern (Fig. 3.8). To determine whether this is the best model, however, the other two options should be explored, comparing the graphs of the data to the models, the correlation, standard error, and residuals (Figs. 3.8–3.12). It may help to use your knowledge of the behavior of logarithmic, exponential, and power functions to begin your investigation. In many cases, the difference is slight for a given domain, and it may be difficult to determine an absolute "best."

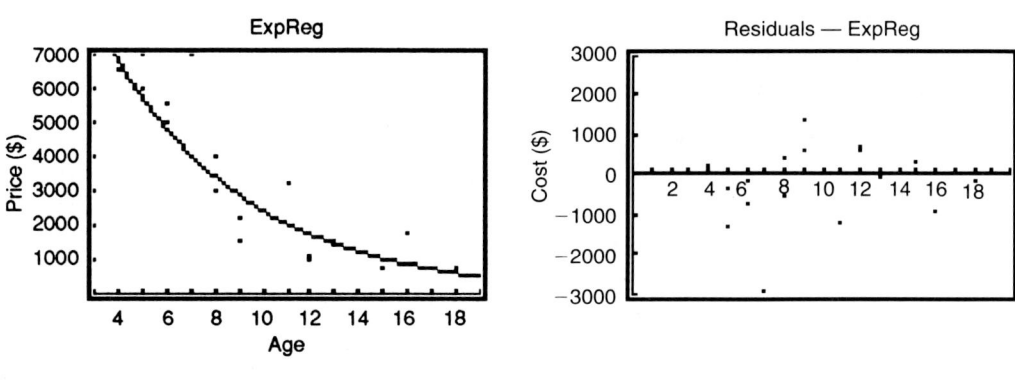

FIGURE 3.9   FIGURE 3.10

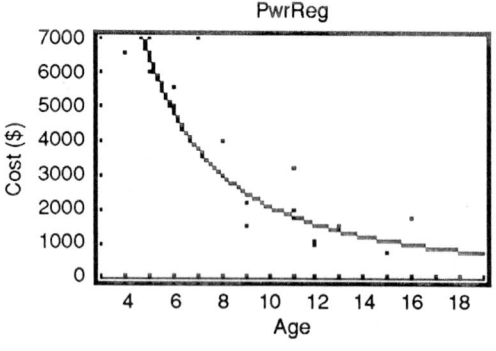

FIGURE 3.11

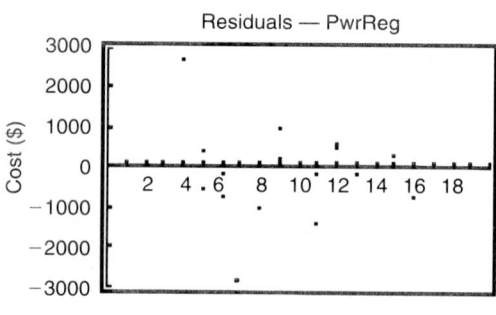

FIGURE 3.12

Equation	Type of regression	a	b	r	Error	RegEq
$y = a + bx$	Lin	7,922	-475	-.856	$1124	$y = 7922 - 475x$
$y = a + b \ln x$	Ln	13,457	-4640	-.898	957	$y = 13{,}457 - 4640 \ln x$
$y = ab^x$	Exp	13,368	.8436	-.889	947	$y = (13{,}368)(.8436^x)$
$y = ax^b$	Pwr	84,975	-1.6	-.898	1131	$y = 84{,}975 x^{-1.6}$

Look at the table and figures. Although the correlation coefficient is -.898, the power regression model has the largest standard error. The residuals for that model are positive for all ages over 12, which means they are not randomly distributed. For older cars, that model would always overestimate the price. The logarithmic and exponential models are close, but the residuals for the exponential regression seem to be more randomly distributed. There are only seven negative residuals for the logarithmic model, and they seem to be farther from the x-axis. The smallest variation in the residuals seems to be in the exponential transformation, although the standard error is only $11 less. Over the given domain, the exponential model seems to have a slight edge.

Exercises

1. The following table gives the fuel economy for selected 1991 cars in city miles per gallon and the average annual fuel cost to operate the cars.

3.4 Nonlinear Models

Car	Miles per gallon	Average annual fuel cost	Car	Miles per gallon	Average annual fuel cost
Honda Civic CRX	49	$375	Honda Civic Wagon	31	$ 568
Nissan NX Coupe	28	585	Ford LTD Wagon	17	938
Geo Metro XFL	53	341	Lamborghini Diablo	9	1977
Volkswagen	37	450	Porsche 928 S4	13	1451
Chrysler Acclaim	24	669	Rolls Royce		
Saab 9000	20	853	Continental	10	1977
GM Caprice			BMW M5	11	1553
Wagon	26	986	Rolls Royce Silver	10	1977
Volvo 240 Wagon	32	782	Mercedes 300TE	16	1279
GM Road Master	16	986	Toyota Camry Wagon	18	892
Cadillac Brougham	16	986			

Source: *World Almanac*, 1991

a) Plot the fuel cost as a function of the miles per gallon (miles per gallon, fuel cost). Fill out the following chart for the different regression models. Draw a sketch of each curve and of the residuals.

Equation	Type of regression	a	b	r	Error	RegEq
$y = a + bx$	Lin					
$y = a + b \ln x$	Ln					
$y = ab^x$	Exp					
$y = ax^b$	Pwr					

b) Find a model that seems to best describe the data. Be prepared to justify your claim.

c) Use your model to predict the average annual cost to operate a Caprice station wagon which is listed 26 miles to the gallon in city driving.

2. The rate of infant deaths per one thousand live births each year in the United States is given in the following table:

Year	1960	1970	1975	1977	1980	1982	1984	1986	1988
Rate	26.0	20.0	16.1	14.1	12.6	11.5	10.8	10.4	9.9

Source: *Statistical Abstract of the United States*, 1989.

a) In order to simplify entering the data, plot the point for 1960 as (60, 26.0), for example. Find the best model for the data.
b) Use your model to predict the number of infant deaths in 1979.
c) Use your model to predict the number of infant deaths in 1989 and in the year 2000.
d) In which of these predictions would you have the most confidence?

3. The vehicle miles traveled per mile of road and the rate of motor vehicle accident deaths per 100,000 population for western states are given in the following table.

State	Miles of travel per mile of road	Deaths per 100,000
Arkansas	228	26.3
Louisiana	514	21.9
Oklahoma	278	21.9
Texas	519	22.2
Montana	107	29.4
Idaho	109	27.3
Wyoming	139	32.1
Colorado	346	20.1
New Mexico	246	36.5
Arizona	294	32.7
Utah	243	22.2
Nevada	180	28.3
Washington	447	17.1
Oregon	240	24.0
California	1227	20.5
Alaska	294	18.4
Hawaii	1750	11.7

Source: *Statistical Abstract of the United States*, 1990

a) Explain why the rate for the number of miles traveled in Hawaii is so large.
b) Plot the data as (miles, death rate), and find the best model for the data. Justify your choice.
c) Use your model to predict the number of deaths in North Dakota, which has a rate of 65 miles traveled per mile of road.
d) The actual death rate in North Dakota was 17.7 deaths per 100,000 people. How does this relate to the predicted rate? What might explain any difference?
e) Predict the number of deaths in Maryland, which has 1271 vehicle miles traveled per mile of road.

4. The pupil teacher ratio in the state of Wisconsin is given along with the average teacher salary for selected years since 1972.

Year	Average Wisconsin teacher salaries	Pupil/teacher ratio	National average teacher salaries
1972	$10,016	22.3	$ 9,705
1980	16,006	17.8	15,970
1982	19,387	17.2	19,274
1986	26,347	16.5	25,201
1987	27,815	16.3	26,556
1988	29,122	16.2	28,008
1989	31,046	17.4	29,567

Source: *Milwaukee Sentinel,* May 3, 1990

a) Plot the pupil teacher ratio as the x and the average teacher salary in Wisconsin as the y. Analyze the data and find the best model.
b) Using your model, describe the relation between the pupil/teacher ratio and the average teacher salary in Wisconsin.
c) Plot the ordered pairs (National salary, Wisconsin salary), and find the best model to describe the relation between the two.
d) Use your model to predict the average salary in Wisconsin if the national average is $33,000.

3.5 Curve Fitting (Optional)

In Section 3.4, different models were analyzed to see how well they "fit," or summarized, the data. How do these models work, and what technique is used to fit the curve? In order to find a good model, the data must be "straightened" by some transformation, and a least squares linear regression line then applied to the transformed data. Consider once again the prices of the used Ford Mustangs. The data are given in the following table:

Age (years)	Price	Age (years)	Price
4	$6500	9	$2250
5	6000	11	2000
5	7000	11	3250
6	4750	12	1000
6	5000	12	1100
6	5500	13	1500
7	7000	15	750
8	4000	16	1750
8	3000	18	750
9	1500		

Remember that the least squares linear regression line had a pattern in the residuals, a relatively large standard error, and would mean that the rate of change or depreciation of the price of a car over time would be constant. To find a better model, a transformation involving logarithms can be applied to the data, and the transformed data analyzed to determine how well a linear relationship will model the data. Several options exist: Take the logarithm of the prices, of the ages, or of both the prices and ages. (There are many other possible models, but the focus here and with the calculator is on these three.) The task is to determine under which of the options will the transformed data come closest to making a straight line.

To transform the data, enter the TRANSFORMATION program. The script at the beginning contains an input code to enable you to take the logarithm or exponential for different sets of variables. To run the program, after the first prompt, enter the number of data points. The next prompt will ask you to choose a transformation: 1 will take the logarithms of the x-variables, 2 will take the logarithm of the y-variables, 3 will take the logarithms of the x- and y-variables, 4 will apply the exponential transformation to the x-variables, 5 to the y-variables, and 6 to both of the variables.

3.5 Curve Fitting (Optional)

TRANSFORMATION PROGRAM

Enter	Display
PRGM EDIT [T] [R] [A] [N] [S] [F] [O] [R]] [M] [A] [T] [I] [O] [N]	:Prgm 1:TRANSFORMATION
PRGM I/O <Input> [ALPHA] [N]	:Input N
PRGM I/O <Disp> [2nd] [ALPHA] " [E][N] [T][E][R][-] [ALPHA] 1 [2nd] [ALPHA] [-][L][N][X][,][-] [ALPHA] 2 [2nd] [ALPHA] [-][L][N][Y][,][-] [ALPHA] 3 [2nd] [ALPHA] [-][L][N][X][,][Y][,][-] [ALPHA] 4 [2nd] ALPHA [-][E][X][P][X][[,][-] [ALPHA] 5 [2nd] [ALPHA] [-][E][X][P] [Y][,][-] [ALPHA] 6 [2nd] [ALPHA] [-][E][X][P][X] [,][Y] ["]	:Disp "ENTER 1 L :NX, 2 LNY, 3 LN :X,Y, 4 EXP x, 5 E :XPY, 6 EXP x,y"
PRGM I/O <Input> [ALPHA] [T]	:Input T
1 [STO] [I]	:1 $\to$ I
PRGM <Lbl> 1	:Lbl 1
PRGM <If> [ALPHA] [T] [2nd] [MATH] <=> 1	:If T = 1
LN [2nd] {x} [ALPHA] [I]) [STO] [2nd] [{x}] [ALPHA] [I])	:ln {x} (I) $\to$ {x} (I)
PRGM <If> [ALPHA] [T] [2nd] [MATH] <=> 2	:If T = 2
LN [2nd] [{y}] [ALPHA] [I]) [STO] [2nd] [{y}] [ALPHA] [I])	:ln {y} (I) $\to$ {y} (I)
PRGM <If> [ALPHA] [T] [2nd] [MATH] <=> 3	:If T = 3
LN [2nd] [{x}] [ALPHA] [I]) [STO] [2nd] [{x}] [ALPHA] [I])	:ln {x} (I) $\to$ {x} (I)
PRGM <If> [ALPHA] [T] [2nd] [MATH] <=> 3	:If T =3
LN [2nd] [{y}] [ALPHA] [I]) [STO] [2nd] [{y}] [ALPHA] [I])	:ln {y} (I) $\to$ {y} (I)

TRANSFORMATION PROGRAM (*continued*)

Enter	Display
PRGM <If> [ALPHA] [T] [2nd] [MATH] <=> 4	:If T = 4
[2nd] [LN] [2nd] [{x}] [ALPHA] [I]) [STO] [2nd] [{x}] [ALPHA] [I])	:e^{x} (I) → {x} (I)
PRGM <If> [ALPHA] [T] [2nd] [MATH] <=> 5	:If T = 5
[2nd] [LN] [2nd] [{y}] [ALPHA] [I]) [STO] [2nd] [{y}] [ALPHA] [I])	:e^{y} (I) → {y} (I)
PRGM <If> [ALPHA] [T] [2nd] [MATH] <=> 6	:If T = 6
[2nd] [LN] [2nd] [{x}] [ALPHA] [I]) [STO] [2nd] [{x}] [ALPHA] [I])	:e^{x} (I) → {x} (I)
PRGM <If> [ALPHA] [T] [2nd] [MATH] <=> 6	:If T = 6
[2nd] [LN] [2nd] [{y}] [ALPHA] [I]) [STO] [2nd] [{y}] [ALPHA] [I])	:e^{y} (I) → {y} (I)
PRGM <If> [ALPHA] [I] [2nd] [MATH] <=> [ALPHA] [N]	:If I = N
PRGM <GoTo> 2	:GoTo 2
[ALPHA] [I] + 1 [STO] [I]	:I + 1 → I
PRGM <GoTo> 1	:GoTo 1
PRGM <Lbl> 2	:Lbl 2

To determine whether $(x, \ln y)$ is linear, enter the data and run the program by entering 19 (the number of data points) for the first prompt and selecting 2 $(x, \ln y)$ for the second prompt. After you have run the program, graph the results in [0, 20] by [6.5, 9] with xscl = 2 and yscl = .2. Calculate the least squares linear regression line, $y = 9.5 - .17x$, and the correlation coefficient of $-.8895$. Use the draw function to graph the line on the points. Paste the regression line onto Y1, and lock it by putting the cursor over the = and pressing [ENTER]. Draw the scatterplot ([2nd] [STAT] DRAW <Scatter>), then select [2nd] [PRGM] <DrawF> [2nd] [VARS], <Y1> [ENTER], and the line will be graphed through the points. How well does the line reflect the points? The scatter becomes larger as x increases, which means there will be a larger error for large x.

Run the ERROR program to find a standard error of 54.67. The relative size of the standard error is difficult to perceive because it is a logarithm. Change the viewing rectangle, and graph the residuals in [0, 20] by [-1, 1], xscl = 2, yscl = .2. The residuals do not show a pattern but do look as if the scatter is larger for large x. How does the linear least squares regression line for the transformed data compare with the original? The correlation for the regression line for (age, ln price), -.89, is slightly stronger than the correlation for (age, price), -.86. The standard errors are difficult to compare because one is a logarithm. The residuals have no pattern, although they are smaller for small ages. The transformation, $(x, \ln y)$, seems to be a better model than using the least squares regression line on the given data.

To retrieve your original data, rerun the ERROR program, and then run the TRANSFORMATION program and select option 5 (EXP Y).

When we explored the nonlinear models in Section 3.4, we looked at three different models. Which of the models compares with the equation

$$y = 9.500625701 - .1700480351x$$

generated by the transformation $(x, \ln y)$? Remember that y is really a logarithm, so the equation should be

$$\ln y = 9.500625701 - .1700480351x.$$

Using the definition of logarithms,

$$e^{9.500625701 - .1700480351x} = y$$
$$e^{9.500625701} e^{-.1700480351x} = y$$
$$13{,}368.08864 e^{-.1700480351x} = y$$

Therefore,

$$y = 13{,}368 e^{-.17x}.$$

And,

$$y = 13{,}368(e^{-.17})^x$$
$$y = 13{,}368(.8436)^x$$

The equation generated by the transformation $(x, \ln y)$ is equivalent to the exponential regression model, which gave as an equation $y = 13{,}368(.8436^x)$. The following exercises investigate the other two options.

Chapter 3 Curve Fitting

Exercises

1. Use a process similiar to the example and analyze the question: Does ($\ln x, y$) yield a better linear relation than ($x, \ln y$) to relate the age and price of used cars? Be sure to look at the graphs of the transformed data and use them in your analysis. How does the equation relate to the logarithmic equation from Section 3.4?

2. Analyze the transformation ($\ln x, \ln y$) with respect to the same question.

3. Justify the fact that $\ln y = b \ln x + a$ is equivalent to $y = kx^b$ where $e^a = k$.

4. The following table gives the population of Nevada since 1860:

Year	Population	Year	Population
1860	6,857	1930	91,058
1870	42,491	1940	110,247
1880	62,266	1950	160,083
1890	47,355	1960	285,278
1900	42,335	1970	488,738
1910	81,875	1980	800,508
1920	77,407	1990	1,206,152

Source: U.S. Bureau of the Census

a) Decide which logarithmic transformation will best "straighten" the data, and find the appropriate least squares regression line.

b) Use your model to predict the population of the state of Nevada in 2000.

c) Challenge: Use your model to find the rate of change of the population of Nevada. (Some knowledge of calculus might help.) What was the rate of change in 1980?

4
Simulation

4.1 Simulation—Probability Equal to 1/2

A method often used to answer complex probability questions is to design and run a simulation. **Simulation** is a procedure to create a model that behaves like the real situation. When the simulation is run a large number of times, data are collected, and based on the experimental results, an estimated probability can be calculated.

Consider the following example: Suppose a person is taking a five-question True/False quiz and makes a random guess for each question. What is the probability of correctly answering at least 3 out of the 5 questions?

In designing a simulation of this problem, many models could be used to simulate the simple event of getting a question right or wrong, but any model that is selected must simulate a probability of 1/2.

One way to do this would be to flip a coin five times, letting "heads" represent a correct answer and "tails" represent a wrong answer. While this is fine for a few times, it is often more effective to use technology such as the TI-81 to perform the simulation. To do this, use the **Rand** (random number) function.

When the Rand function is selected, the TI-81 performs a computation involving a number stored in its memory. This number, called the "seed," is operated on and then generates a random number between 0 and 1 (possibly 0 but never 1). This seed is always the same number and will always generate the same random number sequence. To give the random number generator a seed of your own choosing, you must store an integer seed value in Rand. For example, to store 2 to Rand enter

2 [STO] [ALPHA] [MATH] PRB <Rand> [ENTER].

Note: If you store 0 to Rand, the TI-81 uses the factory set seed value.

To simulate a probability of 1/2, a decimal less than 0.5 will represent a correct answer and a decimal greater than or equal to 0.5 will represent a wrong answer. Since there are five questions on the quiz, five random numbers must be generated.

Chapter 4 Simulation

Enter the following steps to generate five random numbers between 0 and 1:

[MATH] PRB <Rand> [ENTER] [ENTER] [ENTER]
[ENTER] [ENTER] [ENTER]

Figure 4.1 shows five possible random numbers and how each is interpreted.

This trial would be interpreted as having two correct answers. In order to obtain a close estimation for the probability, the simulation should be run a large number of times. Following is the beginning of a chart that can be used to collect the results from each run of the simulation. The chart shows a tally for two correct answers from the first run of the simulation.

Number of correct answers	Tally	Total occurrences
0		
1		
2	|	1
3		
4		
5		

Remember each run of the simulation consists of generating 5 random numbers, interpreting these numbers to find the number of correct answers, and entering the number correct on the chart.

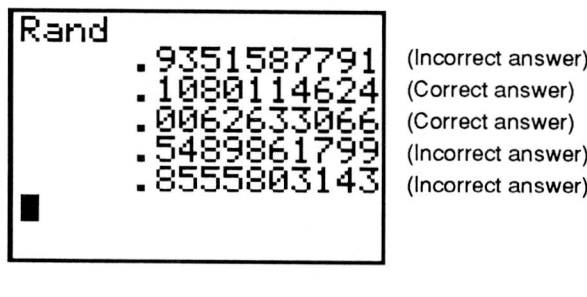

FIGURE 4.1

4.1 Simulation—Probability Equal to 1/2

Here are the results of 30 trials:

Number of correct answers	Tally	Tally occurrences
0	|	1
1	|||||||	7
2	|||||||||	9
3	|||||||	7
4	||||	4
5	||	2
		30 trials

The chart contains the total times out of the 30 simulations that a certain number of correct answers occurred. For example, in 9 of the simulations exactly 2 of the 5 questions were answered correctly.

Using the results of these 30 trials, the estimated probability of answering 3 or more correctly is equal to (7 + 4 + 2)/30, or .433.

As in Chapter One, a histogram can be useful in understanding the data. For this set of statistical data, use the Stat screen, enter the x value as the number of correct answers, and the y value as the number of occurrences. Figure 4.2 shows the histogram of the data in the viewing window [0, 6] by [-1, 10], xscl = 1, yscl = 1. From the histogram, you can estimate the average number of questions that would be answered correctly by guessing. To check your estimate of the mean number of correct guesses, the 1-var screen is displayed in Fig. 4.3. The mean is 2.4. Because the data are from a sample, the standard deviation used is S_x, which is 1.28. An "average" range for the number of correct guesses is 2.4 ± 1.28, or from 1.12 to 3.68, or from about 1 to 4. This is illustrated by the shape of the histogram.

For a more detailed explanation of simulation refer to *The Art and Techniques of Simulation* by Gnanadesikan, Scheaffer, and Swift.

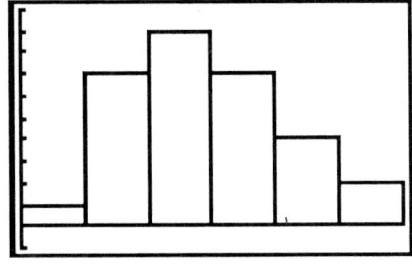

FIGURE 4.2

FIGURE 4.3

Exercises

1. A beginning tennis player's serve is good only 50% of the time. What is the probability that the player's serve will be good at least 4 out of 7 serves? Design a simulation and run it at least 20 times. Show the results in a chart similar to the one shown.

Number of good serves	Tally	Total number
0		
1		
2		
3		
4		
5		
6		
7		

   a) Based on your simulation, estimate the probability that the player's serve will be good at least 4 times.

   b) Sketch the histogram of the data from your simulation. What range values did you use to display the histogram? Describe the shape of the histogram.

   c) Find the mean and the standard deviation of the data from your simulation and display the mean on the histogram. Estimate the percent of the good serves within one standard deviation of the mean.

   d) Find the median of your data, and display the median on the histogram. How do the mean and the median compare?

2. The NBA Championship is determined by a best of seven series; that is, the first team to win 4 games is the winner. If two teams are evenly matched, what is the probability the series will last only 4 games? Design a simulation, run it at least 30 times, and show the results in a chart similar to the one that follows. When running this simulation you are concerned with how many games are needed so one of the teams will win 4 games, not with who won or lost.

Number of games played	Tally	Total number
4		
5		
6		
7		

a) Based on your simulation, estimate the probability that the series will last only 4 games.

b) Draw a sketch of the histogram of the data from your simulation, and describe the shape of the distribution. Estimate a range for the average number of games that must be played to complete the series.

c) Find the mean length of a series and display the mean on the histogram.

d) Find the median length of a series and display the median on the histogram. How do the mean and the median compare?

3. Two people are playing the following game: Three chips are used, one chip with an X on one side and a Y on the other; one chip with a Y on one side and a Z on the other; and one chip with a Z on one side and a T on the other. All three chips are flipped; Player A wins if any two chips match. What is the probability that Player A wins? Design a simulation; run it at least 30 times.

Based on your simulation, what is the probability that Player A will win this game? Would you consider this to be a fair game? Explain.

4. Generate a random number. If the random number is less than 0.5, move one space up on the following diagram. If the random number is greater than or equal to 0.5 then move down on the diagram.

a) How many times do you expect to return to the middle line in randomly selecting 20 numbers?

b) Run the simulation and keep track of the times that you returned to the middle line.

66   Chapter 4   Simulation

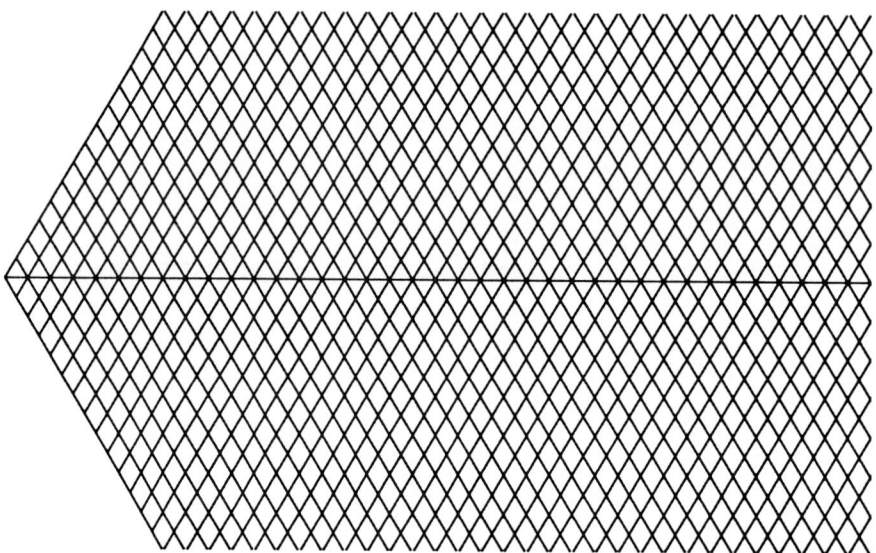

4.2 Probability Not Equal to 1/2

In the examples discussed in the last section each of the outcomes had an equal chance of happening. Suppose the probability of the success of a simple event is not 1/2. Then a new model to simulate the correct probability is needed.

If the probability of the success of an event is 70%, a simulation can be designed to use the numbers from 0 to 9, with 0 to 6 representing a success and 7 to 9 as a failure. Since Rand selects a number from 0 to .9999999999, this random decimal can be awkward to work with. To convert the random decimal to an integer from 0 to 9, the range of the numbers needs to be expanded, and the **Int** function must be used. The expression Int(10*Rand) will accomplish the goal of converting the random decimal to an integer from 0 to 9.

The expression 10*Rand will go from 0 to 9.9999999999, and Int(10*Rand) will give integers from 0 to 9. Thus, to convert the random decimal to an integer from 0 to 9, enter the following statement:

Int(10*Rand).

Use the following steps:

$\boxed{\text{MATH}}$ NUM <Int> $\boxed{(}$ 10 $\boxed{\times}$ $\boxed{\text{MATH}}$ PROB <Rand> $\boxed{)}$ $\boxed{\text{ENTER}}$.

When $\boxed{\text{ENTER}}$ is pressed, a random number from 0 to 9 will be printed.

4.2 Probability Not Equal to 1/2

To generate a random integer from 1 to 10, a 1 needs to be added to the statement as shown:

Int(10*Rand) + 1.

In general, the statement Int($(B+1)$*Rand) + A will produce a random integer from A to $A + B$. In the last example A is equal to 1, and B is equal to 9.

Consider the following example: A multiple choice test consists of six questions, and each has four possible answers, only one of which is correct. If a person makes random guesses on all the questions, estimate the probability of getting at least three questions correct.

In this simulation, the probability of a success is 25%. One model that could be used to simulate 25% is to generate random integers from 0 to 3, with 0 representing a correct answer and 1, 2, and 3 wrong answers. Since there are six questions, six random integers should be generated to complete a trial. To generate random integers from 0 to 3 from the preceding formula, A would equal 0 and B would equal 3. Enter the following statement: Int(4*Rand); and press ENTER six times. A sample run might produce the numbers 3, 2, 1, 3, 1, and 1. This run would mean for this one trial, all the questions were answered incorrectly. Here is a table of data for 30 trials of this simulation:

Number of correct answers	Tally	Total
0	|||	3
1	||||||||||||||||	16
2	||||||	6
3	|||	3
4	||	2
5		0
6		0

Using the results of these 30 trials, the estimated probability of answering at least 3 correctly is 5/30, or about 0.17.

Enter the results as statistical data with the x values as the number of correct answers and the y values as the totals. The 1-var statistics are shown in Fig. 4.4. The average number of correct responses is equal to 1.5. This means you would expect someone to have 1 or 2 questions right out of 6 just by guessing. Using the standard deviation, an expected range of correct answers would be from .46 to 2.54, or from about 0 to 3.

A display of the histogram of the data in the viewing window [0, 6] by [-1, 20], xscl = 1, yscl = 2, is shown in Fig. 4.5.

68   Chapter 4   Simulation

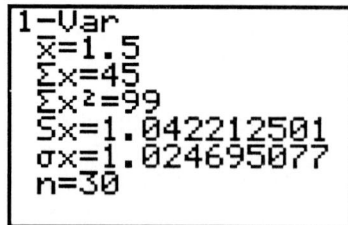

FIGURE 4.4   FIGURE 4.5

Exercises

1. Write a statement that will produce a random integer from:
   a) 0 to 1
   b) 0 to C
   c) 1 to 10
   d) 1 to D

2. A certain baseball player has a .300 batting average. That is, the probability of that player getting a hit for a given time at bat is .300, or 30%. What is the probability that this player will get at least one hit in the next five times at bat? Design a simulation, run it at least 30 times, and show the results in a chart similar to the following one.

Number of hits	Tally	Total
0		
1		
2		
3		
4		
5		

   a) Based on your simulation, estimate the probability the player will get at least one hit in the next five times at bat.
   b) Based on your simulation, estimate the mean number of hits the player will get in the next five times at bat.

c) Make a histogram for the data from your simulation. Explain how your answers in parts (a) and (b) relate to the histogram.

d) Estimate an expected range for the number of hits in five times at bat. Would it be unlikely for the player to have five hits in the next five times at bat?

3. ABC Copter Service flies between Midway Airport and O'Hare Airport using a five-seat helicopter. Past records show only 85% of those who book the trip actually show up. If the airlines sells six tickets for a flight, what is the probability the airlines will have to bump a passenger because of overbooking? Design a simulation, run it at least 30 times, and record your data in the following table.

Number of passengers who arrived	Tally	Total
0		
1		
2		
3		
4		
5		
6		
7		

a) Based on your simulation, estimate the probability that exactly one passenger must be bumped.

b) Find the mean and standard deviation for the number of passengers that arrive for each flight.

c) Sketch a histogram of the data from your simulation. Explain how your answer in part (b) relates to the histogram.

d) Would you recommend that the airline sell more or less tickets for this flight? Explain.

4. A person is standing in the middle of a large plaza. The person randomly takes a step either north, south, east, or west and then stops. From this new position, the person again randomly takes a step either north, south, east, or west. If the person continues this for five steps, how far (straight line distance) would you expect the person to be from the original point after the five steps? Design and run a simulation. You may wish to place the starting position at the origin of the coordinate plane and use the Pythagorean Theorem to find the distance from the origin to the ending point. Remember to record the distance for each trial.

Extension

Write a program that will allow the user to input the lowest and highest random integer needed for the simulation and also the number of trials required for the simulation. The program should print the list of generated random integers.

4.3 Unknown Number of Events

In many simulations the length of a simulation changes from trial to trial. To illustrate this, consider the following problem. A candy manufacturer includes a random letter from the word WIN on the inside of the package. If each letter occurs with equal frequency, what is the mean number of packages of candy a person would have to buy in order to collect the three letters?

To simulate this event, let 0 represent W, 1 represent I, and 2 represent N. Enter the following statement: Int(3*Rand). For this simulation, you must continue to use ENTER until all the digits 0, 1, and 2 have been generated to indicate that at least one W, I, and N have been collected. A sample run might be: 1, 2, 1, 1, 2, 0. In this trial six random numbers were generated before all the digits were printed. A summary of 30 trials is shown in the following table:

Number of packages purchased	Tally	Total
3											9
4							5
5				2
6								6
7							5
8				2
9		0
10			1
		30

The table contains the total number of times a certain number of boxes had to be purchased before the word WIN could be spelled. For example, 9 times out of the 30 only 3 packages had to be purchased.

To find the mean and the standard deviation, enter the results as statistical data with x values from 3 to 10 and the y values as the totals. The 1-var statistics are shown in Fig. 4.6. Because $\bar{x}$ is 5.133, the expected number of packages that must be

4.3 Unknown Number of Events

```
1-Var
x̄=5.133333333
Σx=154
Σx²=900
Sx=1.942861972
σx=1.910206504
n=30
```

FIGURE 4.6

purchased is about 5; and since the standard deviation is about 2, we usually expect to purchase about 3 to 7 packages.

Exercises

1. If the candy manufacturer included a random letter from the word MICKEY on the inside of the package, how many packages would you expect to buy in order to collect all six letters? Assume each letter occurs with equal frequency. Design and run a simulation. You may wish to collect your data in a chart like the one shown here.

Number of packages purchased	Tally	Total
6		
7		
8		
etc.		

   a) Based on your simulation, what is the mean number of packages that had to be purchased in order to collect all six letters?

   b) What is the standard deviation of your data? Find the percent of the data within one standard deviation of the mean. What does this indicate about the number of packages you would expect to buy?

   c) Sketch a histogram of your data, and mark the mean and one standard deviation from either side of the mean on the histogram.

2. About 45% of the American public has O+ blood. If a blood center needs four donors with O+ blood, how many donors, on the average, should they expect before they have received the four pints of O+ blood? Design and run a simulation to estimate your answer.
   a) What assumptions have you made in designing the simulation?
   b) Based on your simulation, how many donors, on the average were necessary in order to get the four pints of O+ blood?
   c) Find the standard deviation for your data, and use it to describe an expected range for the number of donors needed.
   d) Sketch the histogram of the data from the simulation. Mark the mean on the histogram, and also indicate the data that are within one standard deviation of the mean.

3. A basketball player makes 37% of the shots that are attempted. Design and run a simulation. On the average, how many shots must be taken in order to make five baskets?
   a) Based on your simulation, on the average, how many shots must be taken to make five baskets?
   b) What is the standard deviation of the data? What percent of the data is within one standard deviation of the mean?
   c) Sketch a histogram of the data, and mark the mean and one standard deviation on either side of the mean on the histogram. What does this indicate about the number of shots taken?
   d) Compile data from four other students. Based on this compilation, how many shots must be taken to make five baskets? Compare the results of your simulation with the results of the compilation.

5

Probability Distributions

5.1 Methods of Counting

In calculating probabilities, the counting of the total number of outcomes is sometimes difficult because of the large number of possibilities. For example, how many possible arrangements are there for the letters in the word TEXAS? The counting principle states that five different things can be arranged in order $5! = 5 \times 4 \times 3 \times 2 \times 1 = 120$ different ways. In general, n different items can be arranged in order $n! = n(n-1)(n-2) \times \cdots \times 2 \times 1$ different ways. Therefore, the letters in the word TEXAS could be arranged, or permutated, in 5! ways.

To find 5! using your calculator enter the following:

Enter	Display
5 MATH <!> ENTER	5!
	120

The value for n must be an integer between 0 and 69 inclusive.

Sometimes all of the items are not included in the arrangement. For example, how many different arrangements are possible using 2 of the 5 letters in TEXAS? The number of **permutations of r items selected from n items** is

$$_nP_r = \frac{n!}{(n-r)!}.$$

To find $_5P_2$, enter the following keystrokes:

Chapter 5 Probability Distributions

Enter	Display
5 [MATH] PRB <$_nP_r$> 2 [ENTER]	5 $_nP_r$ 2
	20

To find the number of permutations of *n* items, of which *a* items are alike, another *b* items are alike, another *c* items are alike, and so forth, the formula is

$$\frac{n!}{a!\ b!\ c! \cdots}.$$

For example: The number of permutations of letters in the word STATISTICS is

$$\frac{10!}{3!\ 3!\ 2!},$$

since there are 3 S's, 3 T's and 2 I's. Enter the following:

Enter	Display
10 [MATH] <!> [÷] [(] 3 [MATH] <!> [×] 3 [MATH] <!> [×] 2 [MATH] <!> [)] [ENTER]	10!/(3!*3!*2!)
	50400

When *r* items are selected from *n* available items but order is not important, then we want the total number of combinations. The number of **combinations of *r* items selected from *n* items** is

$$_nC_r = \frac{n!}{(n-r)!\ r!}.$$

To find $_5C_2$, enter the following keystrokes:

Enter	Display
5 [MATH] PRB <$_nC_r$> 2 [ENTER]	5 $_nC_r$ 2
	10

Exercises

1. Evaluate the given expressions
   a) $0!$
   b) $\dfrac{50!}{(50-10)!}$
   c) $\dfrac{25!}{(25-6)\,6!}$
   d) $_{10}P_2$
   e) $_{27}C_5$
   f) $\dfrac{_{10}P_8}{_{10}C_8}$

2. If there are 40 floats in a parade,
   a) in how many different ways can all the floats be arranged?
   b) in how many different ways is it possible to select a first place float and a second place float?

3. Find the number of permutations of the letters in the word ARRANGEMENTS.

4. A basketball team consists of 12 players. How many different starting teams (5 people) are possible? (Ignore who's playing which position.)

5. Investigate, using your calculator, whether the following statements are true or false.
   a) $_nP_r = {_nP_{n-r}}$
   b) $_nC_r = {_nC_{n-r}}$
   c) $_nC_0 = {_nP_0}$
   d) $_nC_n = 1$
   e) $_nP_n = n!$

6. What values of n make the following statements true?
   a) $_{n+1}P_3 = {_nP_4}$
   b) $3 \times {_{n+1}C_3} = 7 \times {_nC_2}$

Extension: Approximation for n!

7. When n is large, $n!$ can be approximated using the following formula:

$$n! = 10^k, \text{ where } k = (n + 0.5) \log n + 0.39908993 - 0.4329448n.$$

   a) Use the formula and find the approximation for 50!
   b) Use the factorial key to find 50! and compare the results from part (a).

8. a) Stirling's approximation for $n!$ is

$$n! \approx \sqrt{2\pi n}\, n^n e^{-n}.$$

   Use Stirling's approximation to find 50! How does this compare with the results from Exercise 7?

   b) What is the highest value for $n!$ that you can find using Stirling's approximation?

   c) What is the highest value if you write $n^n e^{-n}$ as $(n/e)^n$?

5.2 Binomial Probability Distribution

Suppose you tossed a coin 20 times. What is the probability that at least half of the tosses will result in heads? How many heads would you expect to see? To answer these questions, simulations similar to those in Chapter 4 could be used. These problems can also be answered by studying the binomial probability distribution. A binomial experiment has the following characteristics:

1. There are a fixed number of trials.
2. Each trial can have only two outcomes.
3. The probability of a success is the same for each trial.

To calculate a binomial probability, we can use the general binomial probability formula:

$$P(x) = {}_nC_x \times p^x \times q^{n-x},$$

where n is the number of trials, p is the probability of a success, q is the probability of a failure, and x is the number of successes.

Consider the following example: A multiple choice test consists of 20 questions; each one has four possible answers, of which one is correct. If all the answers are guesses, find the probability of getting exactly five correct answers.

$$P(5) = {}_{20}C_5 \times (0.25)^5 (0.75)^{15}$$

5.2 Binomial Probability Distribution

To find this probability enter the following keystrokes:

Enter	Display
20 [MATH] PRB <$_nC_r$>	20_nC_r 5 * .25 ^5*
5 [×] .25 [^] 5 [×] .75 [^] 15 [ENTER]	.75^15
	.2023311519

Suppose that the probability is .60 that a person who decides to quit smoking is successful. To find the probabilities that among five persons who decide to quit smoking, 0, 1, 2, 3, 4, or 5 will be successful, enter and run the following program for each of the values 0 to 5.

```
Prgm1: BINOMIAL
:DISP "PROBABILITY"
:INPUT P
:DISP "TRIALS"
:INPUT N
:DISP "SUCCESSES"
:INPUT X
:$N_nC_r$ x * P ^ X * (1–P) ^ (N–X) → Y
:DISP Y
```

Figure 5.1 shows the display for .60 entered as the probability, 5 as the number of trials and 0 as the number of successes.

The following chart shows the probabilities that among the five people 0, 1, 2, 3, 4, or 5 will quit smoking. The second row of the chart shows the probability that exactly 1 person out of 5 will quit smoking is approximately 8%.

78   Chapter 5   Probability Distributions

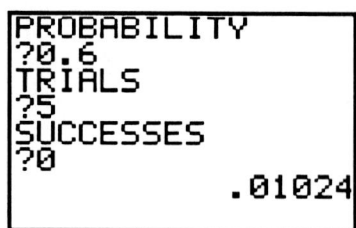

FIGURE 5.1

Number of successes	P(success)	P(success) rounded
0	.01024	1%
1	.0768	8%
2	.2304	23%
3	.3456	35%
4	.2592	26%
5	.07776	8%

To see a pictorial representation of this binomial experiment, enter the results as statistical data with the x values as 0, 1, 2, 3, 4, and 5 and the y values as $P(x)$ rounded to the nearest whole number percent. (To draw a histogram, the y value must be entered as a count or as a whole number.)

Figure 5.2 displays the histogram in the viewing window [0, 6] and [-5, 40] with xscl = 1 and yscl = 5. This histogram is called a **probability histogram,** where the area of each rectangle divided by the total area is the probability of the corresponding value of x.

To further understand the binomial distribution, it is helpful to find the mean and standard deviation of the distribution. Figure 5.3 shows the 1-var statistics for this example. As shown, 3 is the mean number of people that would be expected to quit smoking. According to the data, approximately 84% of the time from 2 to 4 people will quit smoking. It is also true that about 84% of the area in the histogram is from 2 to 4.

5.2 Binomial Probability Distribution

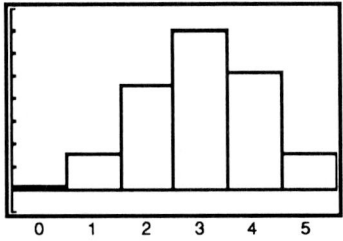

FIGURE 5.2

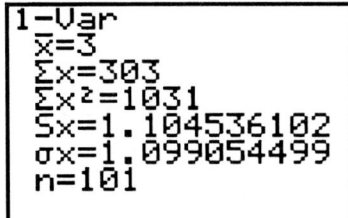

FIGURE 5.3

Exercises

1. Find the $P(x)$ given n, x, and p for each of the following:
    a) $n = 50$, $x = 25$, $p = .5$
    b) $n = 25$, $x = 5$, $p = .10$
    c) $n = 36$, $x = 30$, $p = .75$
2. Women make up about 45% of the labor force.
    a) From 20 randomly selected workers, what is the probability that exactly 6 will be women?
    b) How many women would you expect to have in this random sample?
    c) Design and run a simulation to answer the same questions. How do your answers compare?
3. On a certain stretch of highway, about 65% of all cars are speeding. If a random sample of 6 cars is taken, use the program BINOMIAL and find the probability that 0, 1, 2, 3, 4, 5, or 6 cars are speeding.
    a) Draw a histogram of the probability distribution.
    b) Find the mean and standard deviation of the probability distribution using the 1-var STAT screen.
    c) Mark the mean and one standard deviation from the mean on the histogram. Find the percent of the distribution within that interval.
    d) If a sample of 6 cars was taken, and 1 car was found speeding, would you think this is a likely occurrence? Explain.
4. a) Using the BINOMIAL program, complete the following table for $n = 10$, and $p = .10$.

80   Chapter 5   Probability Distributions

Number of successes	P(success)	P(success) rounded
0		
1		
2		
3		
4		
5		
6		
7		
8		
9		
10		

    b) Display the histogram, and find the mean and median of the distribution.

    c) Find the percent of the total distribution less than or equal to the mean.

    d) What is the probability of at most one success?

    e) What is the probability of at least four successes?

5. A high school basketball player is a 42% free-throw shooter. Use the BINOMIAL program and find the probability that the player will make at least 3 of the next 8 shots. (Assume that each shot is independent of the other shots.)

    a) How many shots would you expect the player to make out of the 8 shots?

    b) Is a likely occurrence that the player makes all 8 shots? Explain.

5.3 Discrete Probability Distributions

In Section 5.2 we investigated the binomial distribution, one of the most important discrete probability distributions. In a discrete distribution, the set of possible values is a finite set or countable set of values. In this section we will study two other discrete probability distributions, the hypergeometric and the Poisson distributions.

    Consider the following example: Four people are randomly selected (without replacement) from a population of 7 men and 5 women. We want to find the probability of getting 3 men and 1 woman.

    There are $_7C_3$ ways to choose the men, since order doesn't matter. For each of these ways, there are $_5C_1$ ways to choose the woman. There are $_{12}C_4$ ways to choose the four people. Thus, the probability is given by

$$P(3 \text{ men}) = \frac{_7C_3 \times {_5C_1}}{_{12}C_4}.$$

To compute using your calculator, enter the following:

Enter	Display
7 [MATH] PRB <$_nC_r$> 3 [×]	7 $_nC_r$ 3 * 5 $_nC_r$ 1/
5 [MATH] PRB <$_nC_r$> 1 [+]	12 $_nC_r$ 4
12 [MATH] PRB <$_nC_r$> 4 [ENTER]	.3535353535

In general, if k objects are chosen without replacement from a set of n objects of which a objects are of one type, then the probability of getting x objects of this type is given by

$$P(x) = \frac{_aC_x \times {_{n-a}C_{k-x}}}{_nC_k}.$$

In the previous example, $n = 12$, $k = 4$, $a = 7$, and $x = 3$.

Here is another example: A deck of cards has 52 cards, of which 12 are picture cards. Five cards are dealt. We want to find the probability exactly three are picture cards. The answer is

$$P = \frac{_{12}C_3 \times {_{40}C_2}}{_{52}C_5} \approx .066,$$

with $n = 52$, $k = 5$, $a = 12$, and $x = 3$.

The following table contains the probabilities of selecting various numbers of men and women from the finite population in the first example.

Chapter 5 Probability Distributions

Number of men	Number of women	Formula	$P(x)$
0	4	$P(0) = \dfrac{_7C_0 * {_5C_4}}{_{12}C_4} = .0101$	$\approx 1\%$
1	3	$P(1) = \dfrac{_7C_1 * {_5C_3}}{_{12}C_4} = .1414$	$\approx 14\%$
2	2	$P(2) = \dfrac{_7C_2 * {_5C_2}}{_{12}C_4} = .4242$	$\approx 42\%$
3	1	$P(3) = \dfrac{_7C_3 * {_5C_1}}{_{12}C_4} = .3535$	$\approx 35\%$
4	0	$P(4) = \dfrac{_7C_4 * {_5C_0}}{_{12}C_4} = .0707$	$\approx 7\%$

To see a pictorial representation of this distribution (called a **hypergeometric distribution**), enter the results as statistical data with the x values as 0, 1, 2, 3, and 4 and the y values as the $P(x)$ rounded to the nearest whole number percent. Figure 5.4 displays a sketch of the histogram in the viewing window [0, 5] by [-5, 50], with xscl = 1, yscl = 5. From the histogram we can estimate the mean number of men, about 2.3, that would be selected, and can see the spread of the data about the mean. Figure 5.5 displays the 1-var statistics for this example.

Another discrete probability distribution is the Poisson distribution. This distribution is often used as a model to describe the probability distribution of such events as the arrival times of customers at a restaurant, the number of calls received at a

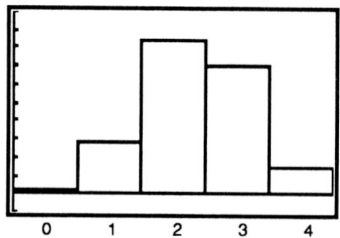

FIGURE 5.4

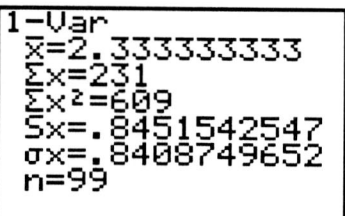

FIGURE 5.5

switchboard, and the arrivals of cars at a gas station. In any Poisson distribution, the probability of getting x successes is

$$\frac{\mu^x e^{-\mu}}{x!},$$

where μ is the mean number of successes during a given time interval.

Consider the following example: Telephone calls come into a certain hotel switchboard at a rate of four calls per minute. Find the probability of receiving exactly two calls in a one-minute interval.

Here $\mu = 4$ and $x = 2$:

$$P(2) = \frac{4^2 \times e^{-4}}{2!}.$$

Enter	Display
4 [x²] [×] [eˣ] [(-)] 4 [+] 2 [MATH] <!> [ENTER]	$4^2 * e\wedge -4/2!$
	0.1465251111

Thus, the probability of receiving exactly 2 calls would be approximately 15%.

Exercises

1. Suppose there are 20 pens in a bag, of which 5 are defective.
   a) If a random sample of 10 pens is taken without replacement, find the probability of getting 5 good and 5 defective pens.
   b) Find the probability none are defective.
2. Suppose there are 40 people, of which 15 are women. If 12 people are chosen for a jury, find the probability of 10 men and 2 women.
3. Among 65 employees of a company, 40 are hourly paid employees.
   a) If 10 employees are chosen without replacement, find the probability that 5 of them are paid hourly.
   b) If the 10 employees are chosen with replacement, find the probability that 5 of them are paid hourly.
   c) If 4 employees are chosen without replacement, find the probability that 2 of them are paid hourly.

d) If the 4 employees are chosen with replacement, find the probability that 2 of them are paid hourly.

4. A bag contains 20 marbles, of which 8 are blue and the rest are red. The following table contains the probabilities of selecting a certain number of blue marbles when 10 marbles are randomly selected without replacement.

x	$P(x)$
0	0%
1	1%
2	8%
3	24%
4	35%
5	24%
6	8%
7	1%
8	0%

a) Display the histogram, and estimate the mean and median of the distribution.

b) What percent of the data is within one standard deviation of the mean?

c) Do you think that selecting 2 blue marbles is a likely outcome? Explain.

5. A service station has a mean number of 5.2 cars that fill up with gas during a certain one-hour interval. Use the Poisson probability formula and complete the following table. Type in the formula for 0 cars. Press enter, then use ▲ to return to the formula and ◄ to change the value of the number of successes.

5.3 Discrete Probability Distributions

Number of successes	P(success)	P(success) rounded
0		
1		
2		
3		
4		
5		
6		
7		
8		
9		
10		
11		

a) What is the probability that exactly 5 cars will fill up with gas during the one-hour interval?

b) What is the probability that at most 2 cars will fill up with gas during the one-hour interval?

c) What is the probability of at least 7 cars filling up with gas during the one-hour interval?

d) Draw a histogram of the probability distribution.

e) Mark the mean and one standard deviation from the mean on the histogram. Find the percent of the distribution within the interval.

f) Why does the 1-var statistics list a mean different from the value of 5.2 given in the problem?

Extension

6. A committee consists of 4 seniors, 7 juniors, 3 sophomores, and 5 freshmen. If 7 people are selected at random, find the probability of getting 2 seniors, 3 juniors, and 2 sophomores.

7. In a Poisson distribution the mean is 4.2. Find the probability of at least 2 successes.

5.4 Normal Distribution

In Sections 5.2 and 5.3 we investigated discrete probability distributions. In this and the next section, we will study the **normal probability distribution.** This is an example of a **continuous** probability distribution, because the random variable has an infinite number of possibilities, and the values cover a continuous interval.

Measurement data (heights, weights, lengths) are usually associated with a continuous random variable and very often with the normal distribution. Figure 5.6 shows the mound shaped distribution of the life of light bulbs. If the life of a bulb were measured with greater and greater accuracy, the distribution could be smoothed to the bell shaped curve shown in Fig. 5.7.

This normal curve can be obtained by graphing the function

$$y = \frac{e^{-(x-\bar{x})^2/2\sigma^2}}{\sigma\sqrt{2\pi}},$$

where $\bar{x}$ is the population mean and σ is the standard deviation of the population.

Figure 5.8 shows the display of the equation entered as the function Y_1 where $\bar{x} = 20$, $\sigma = 5$.

To set the range for a normal curve, use $[\bar{x} - 4\sigma x, \bar{x} + 4\sigma x]$ by $[-.2/D, 1.2/D]$, where $D = \sigma\sqrt{2\pi}$ and the xscl = σ and yscl = .12/D. Thus for $\bar{x} = 20$ and $\sigma x = 5$, use [0, 40] by [-.01, .1], xscl = 5, yscl = .01.

Figure 5.9 shows the results of graphing the Y_1 function.

Change the standard deviation to 10, and enter the new function as y_2. Use the same viewing window.

Figure 5.10 shows the graph of the two normal curves. What effect did the change in standard deviation have on the graph of the normal curve?

Consider the following example: Graph the normal distribution with $\bar{x} = 0$ and $\sigma = 1$ in the viewing window [-4, 4] by [-.1, .5], xscl = 1, yscl = .05. (Before entering this function clear the last two examples.)

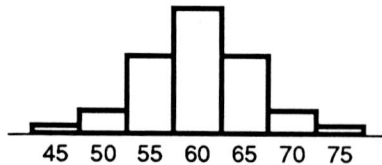

FIGURE 5.6

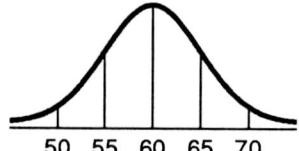

FIGURE 5.7

5.4 Normal Distribution

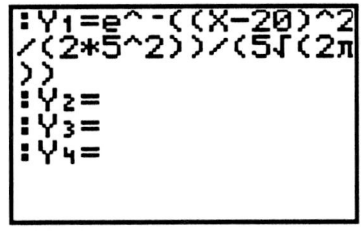

FIGURE 5.8

FIGURE 5.9

In this case, the normal distribution is

$$Y_1 = \frac{e^{-x^2/2}}{\sqrt{2\pi}}.$$

The graph of this normal distribution is shown in Fig. 5.11.

This function is called the **standard normal distribution** and is often used to calculate probabilities of normal distributions. Remember from earlier chapters that probabilities are areas. When finding probabilities associated with the normal distribution, it is useful to shade the area under the curve between the interval of interest. To shade the area under the curve between 0 and 1 use the following keystrokes:

Enter	Display
[2nd] [DRAW] <Shade> (	Shade (0, Y1, 1, 0, 1
Ø [ALPHA] [,] [2nd] [Y-VARS]	)
<Y1> [ALPHA] [,] 1 [ALPHA] [,] Ø	
[ALPHA] [,] 1 [)] [ENTER]	

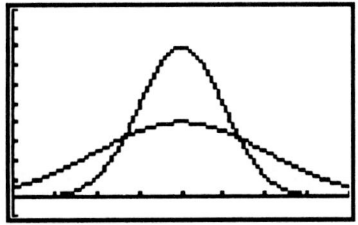

FIGURE 5.10

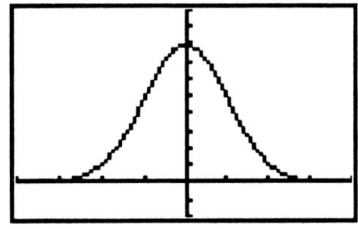

FIGURE 5.11

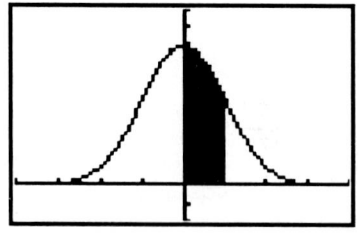

FIGURE 5.12

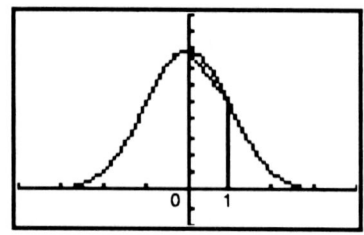

FIGURE 5.13

Figure 5.12 shows the results of the shading. The area of the shaded region can be approximated by finding the area of a trapezoid (see Fig. 5.13). To find the length of the bases of the trapezoid, use the TRACE key and find the y-coordinate when x is Ø and when x is 1. The values are shown in Fig. 5.14. The approximate area of the shaded region would be found by using the formula for the area of a trapezoid, $A = \frac{1}{2}(B + b)h$. Substituting, $A = \frac{1}{2}(.4 + .25)(1) = .325$.

Note: The point on the curve one standard deviation from the mean is the point of inflection for the normal curve.

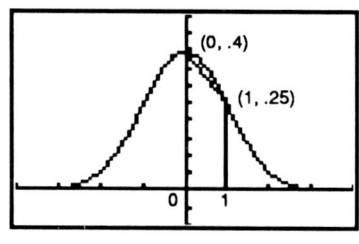

FIGURE 5.14

Exercises

1. Summarize the effect that the standard deviation has on the shape of the normal distribution.

2. a) Graph the normal distribution with $\bar{x} = 10$ and $\sigma = 1$ in the viewing window [6, 24] by [0, 0.5], xscl = 1, yscl = 0.1. Sketch the curve.

b) Graph the normal distribution with $\bar{x} = 20$ and $\sigma = 1$. Sketch the curve and describe the effect the change in the mean has on the normal distribution.
3. Graph the normal distribution with $\bar{x} = 15$ and $\sigma = 5$. List the viewing window that was used. Determine the y values when $x = 10$, 15, and 20.
4. Graph the standard normal distribution, $\bar{x} = 0$ and $\sigma = 1$.
    a) Shade the area under the curve from $x = -1$ to 1. Sketch the results. Estimate the area under the curve from $x = -1$ to 1.
    b) Shade the area under the curve from $x = 0$ to 2. Sketch the results. Estimate the area under the curve from $x = 0$ to 2.
5. Graph the normal distribution with $\bar{x} = 50$ and $\sigma = 10$ in the viewing window [10, 90] by [-.01, .05], xscl = 5, yscl = .01. List the keystrokes that are needed to shade the area under the curve from $x = 35$ to $x = 65$. Sketch the results and estimate the proportion of the distribution that was shaded.
6. If I.Q. scores are normally distributed with a mean of 100 and a standard deviation of 15, graph this normal distribution and shade the area under the curve from $x = 110$ to 120. Sketch the results. Estimate the percent of the total area that is shaded.
7. The weight of all male students at a certain school is normally distributed with a mean weight of 148 pounds and a standard deviation of 15 pounds. Graph this normal distribution, and shade the area under the curve for x between 120 and 145 pounds. Sketch the results and estimate the percent of the total area that is shaded.

5.5 Area Under the Normal Curve

In the last section we investigated the effect the mean and the standard deviation have on the normal curve. Remember that $\bar{x}$ locates the curve on the x-axis, and σ determines the shape of the distribution. As the value of σ increases, the curve becomes flatter and more spread out.

Even though the normal curve can take different shapes, it is a probability distribution. The area under this distribution corresponds to the probability that an x value will fall in a specified interval. Since the areas over intervals represent probabilities, the total area under the curve and above the x-axis equals 1.

In the last section we also found an approximation for the area under the normal curve by finding the area of a trapezoid. The following program will find an approximation for the shaded area, and thus for probabilities. The program uses Simpson's method, found in Program Simpson, to estimate the area under a curve between two bounds. Program Area also will use the program Normal if you wish a graph of the given situation. Note that Program 6 is Simpson and Program 7 is Normal.

Prgm 5: Area
:Disp "MEAN"
:Input U
:Disp "STD DEV"
:Input V
:"e ^ (-(X – U)² / (2V²))
/ (V √ (2π))" → Y1
:Disp "LOWER LIM"
:Input A
:Disp "UPPER LIM"
:Input B
:Prgm 6
:Disp "GRAPH? Ø = NO 1 YES"
:Input T
:If T = Ø
:End
:Prgm 7
:End

Note: Skip these steps if you don't want the graphing program.

The following program uses Simpson's Rule to estimate the area.

Prgm 6: SIMPSON
:All-Off
:2Ø → F
:Ø → S
:(B – A) / 2F → W
:1 → J
:Lbl 1
:A + 2(J – 1) W → L
:A + 2JW → R
:(L + R) / 2 → M
:L → X
:Y1 → L
:M → X
:Y1 → M
:R → X
:Y1 → R
:W(L + 4 M + R) / 3 + S → S

5.5 Area Under the Normal Curve

:IS > (J, F)
:Goto 1
:Disp "AREA IS"
:Int (1000 S + .5) / 1000 → S
:Disp S

An explanation of program 6 can be found on pp. 9–12 of the *TI-81 Guidebook*. The program that follows is used to draw a graph of the situation.

Prgm 7: NORMAL
:Clr Draw
:V √(2π) → D
:U – 4V → Xmin
:U + 4V → Xmax
:-.2 / D → Ymin
:1.2 / D → Ymax
:V → Xscl
:.12/D → Yscl
:DispGraph
:Shade (Ø, Y1, 1, A, B)

Consider the following example: Find the area under the standard normal distribution ($\bar{x} = 0$, $\sigma x = 1$) from $x = -1$ to 1. To find the solution, run the Area program, enter Ø as the mean, 1 as the standard deviation, -1 as the lower limit, and 1 as the upper limit. The screen should display the AREA IS .683.

To see the graph, press 1. Figure 5.15 shows the screen display of the graph. To return to the Home screen, press [2nd] [QUIT].

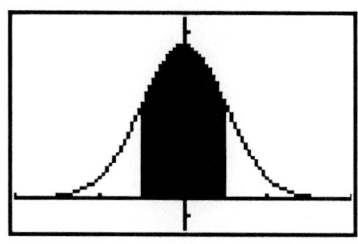

FIGURE 5.15

Consider the following example: Suppose that the weight (lb) of a certain type of fish caught in a lake is normally distributed with a mean value of 3.8 lb and a standard deviation of 1.1 lb. What is the probability that a fish caught will weigh between 3 and 5 lb?

Run the Area program to find the solution. Enter 3.8 for mean, 1.1 for standard deviation, enter 3 as the lower limit, and 5 as the upper limit. The result is shown here:

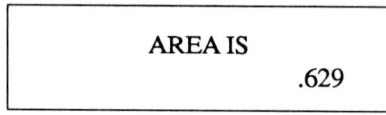

Figure 5.16 shows the screen display of the graph.

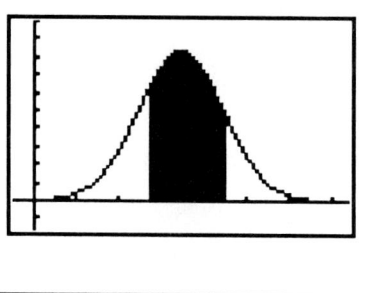

FIGURE 5.16

Exercises

Use the programs in this section to do the exercises.

1. Find the area under the standard normal curve that lies between the given limits. For each, sketch the curve and shade the area to be found.
    a) between $x = 0.8$ and $x = 1.38$
    b) between $x = -.43$ and $x = .62$
    c) to the right of $x = 1.94$
        (enter 4 for the upper limit)
    d) to the left of $x = -1.6$
        (enter -4 as the lower limit)
2. Given that a random variable has a normal distribution with a mean of 10 and a standard deviation of 2, find the probabilities that the variable will take on a value

a) between 7 and 13.
b) between 9 and 12.5.
c) less than 8. What did you enter as the lower limit?
d) greater than 13. What did you enter as the upper limit?

3. Gasoline mileage for a particular type of car is said to be normally distributed with a mean of 18.3 and a standard deviation of 2.3 miles per gallon. If you randomly select one of these cars, what is the probability that the car will get at least 16 miles per gallon?

4. The number of raisins in 71 boxes of raisins is given in the following table. Enter the data.

Number of raisins	Number of boxes	Number of raisins	Number of boxes
26	1	32	13
27	1	33	7
28	7	34	8
29	5	35	4
30	6	36	5
31	8	37	5
		38	1

a) Find the mean and standard deviation, and draw a histogram of the data.
b) What percent of the boxes have from 29 to 35 raisins?
c) Use the Area program and find the probability that a randomly selected box will have from 29 to 35 raisins. [Remember to use the mean and standard deviation from part (a).]
d) Compare the histogram in part (a) to the normal curve in part (c).

Extension: The Normal Distribution Used as an Approximation to the Binomial Distribution

In Section 5.2 we studied the binomial distribution and solved problems similar to the following example: A baseball pitcher throws strikes 40% of the time. What is the probability of his throwing at most 150 strikes of his next 400 pitches?

If we attempt to use the Binomial program from Section 5.2, the calculator will display an error 01 MATH. This error means that some calculation result is greater than 1E100.

To solve this problem, we will use the Area program for finding a shaded area under the normal distribution. Generally, when η is large, the normal distribution can be used to approximate the binomial distribution.

The mean of the binomial distribution is the product of the number of trials and the probability of a success:

$$\mu = \eta p.$$

The standard deviation is the square root of the product of the number of trials, the probability of a success, and the probability of a failure:

$$\sigma = \sqrt{\eta p q}.$$

For the given example, the mean of distribution would be

$$\mu = \eta p = 400(.40) = 160.$$

The standard deviation would be

$$\sigma = \sqrt{\eta p q} = \sqrt{400(.4)(.6)} \approx 9.8.$$

Figure 5.17 shows the normal distribution and the area shaded to correspond to at most 150 strikes in the next 400 pitches. Since we are converting from a binomial (discrete) distribution to the normal (continuous) distribution, 0.5 was added to the value of 150 as a continuity correction.

To calculate the probability of at most 150 strikes in the next 400 pitches, run the Area program and enter 160 for the mean, 9.8 for the standard deviation, 150.5 as the upper bound, and 120 the lower bound. The area is .166. Therefore, the probability of the pitcher throwing at most 150 strikes out of the next 400 pitches is approximately 17%.

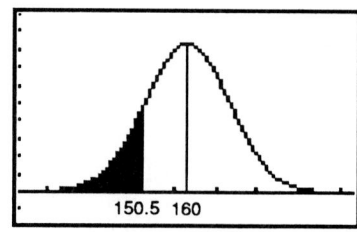

FIGURE 5.17

5. Find the probability of passing a 100 question true-false test by guessing all the answers, if 60% is passing. (*Hint:* $p = .50, q = .50, \eta = 100$.)
   a) Find the mean and standard deviation for this binomial probability.
   b) Use the Area program and find an approximate value for the probability. List the values that you entered for the mean, standard deviation, and upper and lower bound. Sketch the distribution.
6. Past records show that 55% of the freshmen class at a college eventually graduate from that school. Find the probability that between 1550 and 1700 students will graduate from a class of 3000 freshmen.

For a more detailed explanation of the normal distribution as an approximation for the binomial distribution, refer to *Introduction to the Practice of Statistics* by Moore and McCabe.

6

Inference

6.1 Sampling Distribution

A company wants to know what percent of consumers use their product. A TV network would like information on the proportion of the TV viewing public that watches the station's programming. In both of these cases, information is needed from a large number of people. A sample of the population is studied in order to gain information or make inferences about the whole population.

Consider the following example: *American Demographics* magazine stated that about 60% of parents in the United States revealed that their children help to decide what full service restaurant the family chooses for dinner. If a random sample of 20 parents is selected, what is a likely number of parents that would agree with the statement from the *American Demographics* magazine?

One way to answer the question is to design and run a simulation (see Section 4.2). For this example, generate 20 random digits between 0 and 9, with the digits 0 to 5 representing a set of parents that agree with the statement and the digits 6 to 9 representing those parents that disagree.

Enter the following keystrokes:

[MATH] NUM <Int> [ENTER] [(] 10 [×] [MATH] <PRB> <Rand> [)] [ENTER].

Each time the [ENTER] key is used, a random digit from 0 to 9 will be displayed. Here is a sample run of one trial:

1 2 6 0 2 5 8 0 3 2 6 6 9 6 6 8 8 8 6 8.

From this trial, 8 of the 20 numbers were between 0 and 5. This means that 8 out of 20 parents would agree that their children help to decide where the family goes for dinner. If this simulation is run again, it is unlikely that the number of parents that agree would again be 8. Instead we would expect some variability.

Chapter 6 Inference

In the following table are the results of 40 trials, each simulating a survey of 20 sets of parents.

Number of parents who agree	Tally	Frequency	Number of parents who agree	Tally	Frequency
0		0	10	||||	4
1		0	11	|||||||	7
2		0	12	|||||||	7
3		0	13	|||||	5
4		0	14	||||	4
5		0	15	|||	3
6	|	1	16		0
7	|	1	17	|	1
8	||||	4	18		0
*9	|||	3	19		0
			20		0

Practice reading the chart. For example, at the *, in 3 of the trials 9 out of 20 parents agreed. The number of parents who agreed varied from trial to trial (sample to sample), from a low of 6 to a high of 17. Figure 6.1 shows the histogram of the sampling distribution in the graphing window of [5, 20] by [-2, 20], xscl = 2, yscl = 2. Figure 6.2 displays the 1-var statistics of the data. Using 11.4 as the mean and 2.4 as the standard deviation, the range of the data within two standard deviations of the mean is from 6.6 to 16.2, which is about 97% of the data.

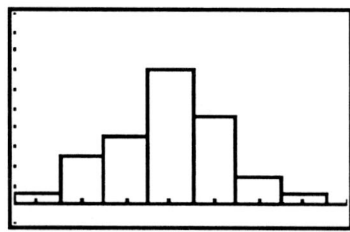

FIGURE 6.1

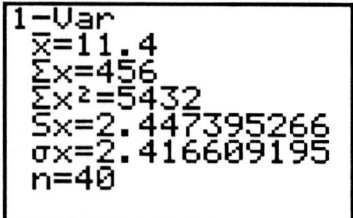

FIGURE 6.2

A box plot can also help summarize the sampling distribution. Here is a 90% box plot of the results of the 40 trials.

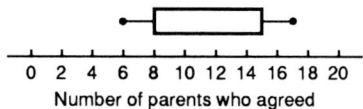

In a 90% box plot, at least 90% of the 40 trials must lie on the edge or within the box. Approximately 5% of the 40 trials lie on the segments on either side of the box (whiskers). Since 90% of 40 is 36, at least 36 of the number of parents who agreed must be in the box, and the remaining 4 are outside. In order to get at least 36 in the box, the edges were drawn at 8 and 15. This resulted in 37 trials in the box, 2 trials on the left whisker and 1 trial on the right whisker.

At least 90% of the sample trials lie between 8 and 15. Any number inside the box is a "likely" event. For example, using the 90% box plot for this distribution, 8 out of 20 randomly selected parents are likely to agree that their children help to select a restaurant. It would be unlikely to find 16 out of 20 randomly selected parents who agreed. From this simulation, we can conclude that 90% of the time if a random sample of 20 parents were asked if their children helped to decide where to eat, anywhere from 8 to 15 parents would agree.

For further discussion of likely/unlikely samples, see *Exploring Surveys and Information from Samples* by Landwehr, Swift, and Watkins.

Exercises

1. The Bureau of Labor Statistics reported that 25% of people with full-time jobs spend 49 hours or more on the job per week. If you take a random sample of 20 people with full-time jobs, how many people are likely to spend 49 or more hours on the job?

   a) Design and run a simulation to answer the question. Run the simulation 10 times, and combine your results with three other students. Display the results in a chart similar to the following:

Chapter 6 Inference

Number of yesses	Tally	Frequency
0		
1		
2		
.		
.		
.		
20		

  b) Draw the histogram of the results.
  c) What is the average number of people in a set of 20 who work more than 49 hours?
  d) What is the standard deviation?
  e) Estimate the percent of the trials that are within two standard deviations of the mean.
  f) Draw a 90% box plot of the data.
  g) Compare the results of parts (e) and (f).
  h) How many people who work full time are likely to spend 49 or more hours on the job?

2. The Marriott Corporation states that about 70% of extended stay travelers (people who stay more than one night at a motel) were under 45 years of age. If a random sample of 20 extended stay travelers is taken, how many travelers are likely to be under 45 years of age?
  a) Design and run a simulation to answer the question. Run the simulation 10 times, and combine your results with three other students. Display the results in a chart similar to the following:

Number under 45	Tally	Frequency
0		
1		
2		
3		
.		
.		
.		
20		

b) Draw a histogram of the results.

c) What is the mean and standard deviation of the number of extended stay travelers that are under 45 years of age?

d) Use the area program from Section 5.5 to estimate the percentage of the trials within two standard deviations of the mean.

e) Draw the 90% box plot of the data.

f) How many extended stay travelers are likely to be under 45 years of age? Compare this answer with the results in part (d).

3. The following chart displays the results of 100 random samples, each of size 20, from a population of 40% yesses.

Number of yesses	Frequency	Number of yesses	Frequency
0	0	10	10
1	0	11	8
2	1	12	3
3	3	13	2
4	7	14	1
5	9	15	0
6	11	16	0
7	10	17	0
8	16	18	0
9	19	19	0
		20	0

a) Draw a histogram of the data.

b) Calculate the mean and standard deviation of the data.

c) Use the area program to estimate the percent of the data within two standard deviations of the mean.

d) Draw a 95% box plot of the data (remember at least 95% of the data must be within the box).

e) What are the likely sample number of yesses?

f) Compare the results from parts (d) and (c).

4. Set up a simulation to answer the following: Sixty percent of the students at a certain high school ride the bus to school. If you take a random sample of 20 high school students from this school, would you be likely to find exactly 7 who ride the bus to school? Explain.

Extension

Work problems 1 and 2 of this section for samples of size 40. What happens to the likely number of sample yesses?

6.2 Sampling Distribution of Sample Means

Many statistical studies make inferences concerning the population mean. These inferences are often based on the sample mean. In the last section we investigated the variation in samples when a population proportion is known. In this section we will study how sampling variability causes the sample mean to differ from one sample to another. Investigating the distribution of the sample means will help to explain some very important ideas in statistics.

We want to generate 10 random digits between 0 and 9 and find the mean of this random sample.

To generate this random list, enter the following keystrokes:

[MATH] NUM <Int> [ENTER] [(] 10 [×] [MATH] PRB <Rand> [)] [ENTER]

The results of one such trial are

5 7 4 1 3 3 5 3 6 7.

The mean of this sample is 4.4. In order to understand how this sample mean can be used to describe the population mean, we will need to increase the number of trials. Instead of entering the same keystrokes, the program that follows generates a random list of digits between 0 and 9, finds the sample mean for each run or trial, and displays a histogram of all of the sample means.

```
Prgm: SAMPLE
:ClrStat
:ClrDraw
:Disp "SAMPLE"
:Disp "SIZE"
:Input S
:Disp "TRIALS"
:Input N
:Ø → J
:Ø → I
:Disp "MEANS"
:Lbl 2
:Ø → T
:Ø → C
:Lbl 1
:Int (10 * Rand) → R
:T + R → T
:C + 1 → C
:If C < S
:Goto 1
:T/S → V
:Disp V
:I + 1 → I
:V → {x}(I)
:J + 1 → J
:If J < N
:Goto 2
:Hist
```

Before running the program, the range for the graphics window must be set. For the following examples the range was [2, 7] by [-2, 20], xscl =.5, yscl = 2. To run the program, enter the sample size and the number of trials when prompted.

Figure 6.3 shows a histogram that was displayed with 10 entered as the sample size and 40 as the number of trials. Use the trace to estimate the mean of the sample means and 1-var to find the calculator mean. The mean of the 40 trials was 4.48 and the sample standard deviation was 0.85.

FIGURE 6.3

Figure 6.4 displays a histogram for a sample size of 20 with 40 as the number of trials. The mean of the sample means for these 40 trials was 4.36, and the standard deviation was .535.

Figure 6.5 shows a histogram with 40 as the sample size and 40 as the number of trials. (*Note:* The σ_x displayed under the 1-var statistics is not the standard deviation of the population. It is the standard deviation of the sample using N as a divisor.) The mean for this set of data was 4.569, and the standard deviation was .407.

The first thing to notice about the histograms is their shape. Each of the three looks reasonably close to a normal curve. The resemblance would even be more normal if each histogram were based on many more than 40 sample mean values. Second, each of the histograms is centered at approximately 4.5. If we assume that each random digit will occur with equal frequency, then the population mean would be 4.5, and the population standard deviation would equal 2.87. Finally, the spread of the histograms changed from one sample size to another. The smaller the sample size, the more the distribution spreads out about the mean.

These three observations lead to some very important rules concerning the sampling distribution of the sample means. First, when the sample size is large

FIGURE 6.4

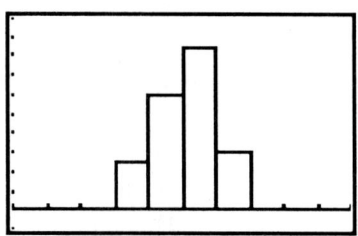

FIGURE 6.5

(generally >30) the sampling distribution is approximately normal with the mean of the sample means, $\mu_{\bar{x}}$, equal to the population mean, μ. Also, as the sample size increases, the sample standard deviation (S_x) decreases. Statisticians have shown that the relationship between the standard deviation of the distribution of sample means and the population standard deviation is

$$S_{\bar{x}} = \frac{\sigma}{\sqrt{n}},$$

where n is the sample size, $S_{\bar{x}}$ is the standard deviation of the sample means, and σ is the standard deviation of the population.

Exercises

1. a) Assuming that the population standard deviation is 2.87, find the standard deviation of the sample means when $n = 10, 20,$ and 40, using the relationship between the population standard deviation and the standard deviation of the sample means.
    b) Compare the results from part (a) with the results of the three simulations in this section.
2. Enter and run the program in this section for a sample size of 80 and 40 trials.
    a) Sketch the histogram.
    b) Find the mean and standard deviation of the sample means.
    c) Use the results from part (b) and estimate the population mean and standard deviation.
    d) In your own words, what are the relationships between the sample statistics and the population statistics?

6.3 Testing a Claim About a Mean

A common problem in statistics is to make decisions about the mean of a population. For example, a local pizza parlor claims that the average delivery time is 23.5 minutes, but every time you order a pizza it seems to take more than 23.5 minutes for your order to arrive. To test the claim that the time is greater than 23.5 minutes, a random sample of 40 delivery times was obtained. The times are listed in the following stem-and-leaf plot.

	Delivery times
18	2
19	2 5
20	6
21	3 6
22	1 3 4 4 7 9
23	0 1 3 4 5 7
24	2 6 7 8 9
25	0 2 2 2 6 8
26	0 1 4 4 5 8 8 9
27	6 7
28	
29	1

Key: 26 | 0 represents 26.0 minutes.

Figure 6.6 shows the 1-var statistics for the 40 times. Do the data provide sufficient evidence to conclude that the delivery time is greater than 23.5 minutes? Does the sample mean of 24.17 represent a statistically significant difference from the claim of 23.5 minutes? If the sample of delivery times had a mean of 30 minutes, we would probably conclude that the stated claim of 23.5 minutes was wrong. However, the difference between 23.5 and the actual sample mean of 24.17 is not quite that obvious.

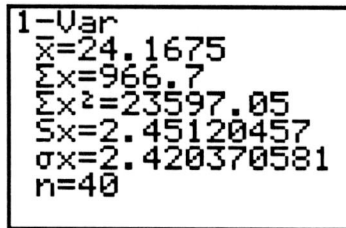

FIGURE 6.6

In the last section, we investigated the properties of sample means. We discovered that sample means tend to be normally distributed with a mean equal to the population mean and a standard deviation equal to

$$\frac{\sigma}{\sqrt{n}}.$$

This information can be used to decide how unusual a sample mean of 24.17 is for a sample of 40 randomly selected times from a population that is assumed to have a mean of 23.5 minutes. If 24.17 minutes occurs often, then we would conclude that the claim of 23.5 minutes is correct and our sample mean was due to sampling variability. But if it is unlikely that a random sample of 40 times will produce a mean of 24.17, then we would disagree with the claim.

To test the claim, find the probability of obtaining a sample mean of 24.17 minutes or more from a population with a mean of 23.5 minutes. Assuming that the sample was taken from a population with a mean of 23.5, the distribution of the sample means will have a mean of 23.5 and a standard deviation of

$$\frac{2.45}{\sqrt{40}} = 0.39.$$

Figure 6.7 shows the distribution and the shaded area corresponding to the probability of obtaining a sample mean of 24.17 or more. *Note:* If the sample size is greater than 30 you may use the value of the standard deviation as an estimate for the population standard deviation.

To find the probability, use the Area program (Section 5.5) and enter 23.5 for the mean, 0.39 for the standard deviation, 24.17 for the lower bound and 25 for the upper bound. The calculated area is 0.041. Therefore, the probability that a population with a mean of 23.5 has a sample mean of 24.17 or more is 4.1%. Our sample mean of 24.17 or more would occur by chance only 4.1% of the time. This makes the claim of a 23.5 minute delivery time seem incorrect. Generally, statisticians use the value of 5% or less as the value that separates a significant difference from chance variability. If an event happens less than 5% of the time, that event is said to be significant at the 5% level.

For a more detailed explanation of testing a claim about a mean see *Introduction to the Practice of Statistics,* Moore and McCabe.

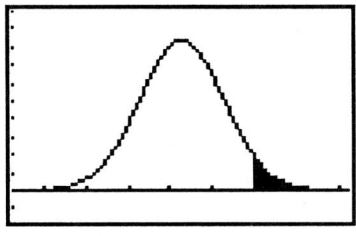

FIGURE 6.7

Exercises

1. On a college entrance exam, the mean score is 510, and the standard deviation is 85. An administrator of a high school claims that the graduates of the high school scored higher than the mean score of 510. A random sample of 55 graduates' test scores showed a mean of 535.

    a) Find the mean and the standard deviation of the sample means.

    b) Test the claim that the graduates did better than the mean score of 510. Use the Area program and find the probability of obtaining a sample mean of 535 or more from a population with a mean of 510. Enter 510 as the mean, enter the standard deviation found in part (a), 535 as the lower bound and 550 (about four standard deviations above the mean) as the upper bound. Interpret the results. Do you believe the administrator's claim? Why or why not?

2. A consumer group claims that the mean miles per gallon of a certain model of car is 25. The following stem-and-leaf plot displays the results of tests done on 36 randomly selected model cars. Test the claim that the mean number of miles per gallon is greater than 25.

Miles per gallon	
16	1
17	
18	5
19	1
20	4
21	3 9
22	5
23	6 6 7 9
24	1 3 7 7
25	3 5 6 9
26	8
27	4 4 6
28	0 1 3 3
29	1 2 5
30	4
31	9 9
32	3 8
33	2

Key: 29 | 1 represents 29.1 miles per gallon.

a) Find the mean and standard deviation for this sample of 36 model cars.

b) Find the mean and standard deviation of the sample means. Assume the population standard deviation is equal to the sample standard deviation found in part (a).

c) Use the Area program and find the probability of obtaining a sample mean greater than or equal to the mean found in part (b), from a population with a mean of 25 miles per gallon. Enter 25 as the mean, enter the standard deviation found in part (b), and enter the sample mean as the lower bound and 28 as the upper bound. Interpret the results. Do you believe that the car gets 25 miles per gallon? Why or why not?

3. The packager of raisins claims that the average box contains 36 raisins. The following stem-and-leaf plot displays the number of raisins found in 30 randomly selected 1/2-oz boxes. Test the claim that the mean number of raisins is less than 36.

Number of raisins	
2	
*	6 7 9
3	1 3 3 3 4 4 4 4
*	5 5 5 5 6 6 7 7 7 7 8 8 8 9 9
4	0 0 1 3

Key: 3 | 1 represents 31 raisins in a 1/2-oz box.

a) Find the mean and standard deviation for this sample of 31 1/2-oz boxes of raisins.

b) Find the mean and the standard deviation of the sample means.

c) Use the Area program and find the probability of obtaining a sample mean less than or equal to the mean found in part (b), from a population with a mean of 36 raisins. Since the shaded area is to the left of the mean, enter 33 as the lower bound and the sample mean as the upper bound. Interpret the results. Do you believe the claim? Why or why not?

4. A local newspaper reported that the average price of a house in town was $60,000. The Realtor took a random sample of 32 homes. The results of the sample follow:

Home prices ($)			
90,480	91,947	74,000	16,645
49,084	42,497	18,053	43,670
65,065	22,238	21,674	58,207
93,035	89,889	84,497	90,523
40,648	42,315	23,068	20,751
77,585	87,035	27,280	50,859
78,058	29,010	52,946	25,819
22,268	43,641	91,743	34,833

Test the claim that the mean price of a home is less than the $60,000 that was reported in the newspaper.

6.4 Chi-Square

A department store stocks blue jeans that are identical except they have been washed by different methods. A random sample of 32 sales showed the following purchases:

Method of wash	Number sold
Glacier washed	5
Stone washed	6
Volcano washed	8
Pepper washed	13
	Total 32

Does this data indicate that one type of washed jeans is preferred over the others, or are consumers buying jeans at random?

6.4 Chi-Square

If the customer has no preference for a given type of jean, we would expect the number sold to equal 8 for each type:

Category	Observed number sold	Expected number sold
Glacier	5	8
Stone	6	8
Volcano	8	8
Pepper	13	8

If the difference between the observed number sold and the expected number sold can be attributed to sampling variation, then there seems to be no preference. On the other hand, if the discrepancy between the observed number and the expected number is large, we could conclude customers do show a preference.

Statisticians use the **chi-square statistic** (pronounced "ki") to measure the difference between observed values and the expected results. The formula for chi-square is

$$\chi^2 = \Sigma(O - E)^2/E,$$

with O = observed value and E = the expected value.

Chi-square for this example would be calculated using the following procedure:

Category	Observed	Expected	$(O-E)$	$(O-E)^2$	$\dfrac{(O-E)^2}{E}$
Glacier	5	8	-3	9	9/8
Stone	6	8	-2	4	4/8
Volcano	8	8	0	0	0/8
Pepper	13	8	5	25	25/8

$$\chi^2 = 9/8 + 4/8 + 0/8 + 25/8 = 4.75$$

To find chi-square on the calculator, enter the following program:

```
Prgm: CHISQU
:Disp "CELLS"
:Input C
:Disp "SAMPLE"
:Input N
:Ø → L
:N/C → E
:1 → Q
:Disp "OBSERVE"
:Lbl 1
:Input O
:(O – E)2 / E → X
:X + L → L
:Q + 1 → Q
:If Q < C
:Goto 1
:Disp "CHI2"
:Disp L
```

When you run this program, enter a 4 after the prompt CELLS, for the number of categories. After the prompt SAMPLE, enter 32 for the sample size; for the prompt OBSERVE enter a 5,6,8, and 13 after each successive question mark. The display should read

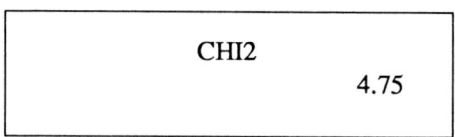

The next section will investigate how to interpret this value of chi-square.

Exercises

1. A die was rolled 36 times with the following results:

Outcome	Frequency	Expected frequency
1	8	
2	5	
3	9	
4	6	
5	3	
6	5	
	36	

   Complete the expected frequency column in the chart, and use the Chi-square program to find the value of chi-square.

2. A high school statistics class conducted a soft drink taste test. In the following chart are the data that were collected. If each soft drink is equally likely to be chosen, find the expected number and find the value of chi-square.

Type of soft drink	Number observed	Number expected
Coke	5	
Pepsi	6	
7-Up	9	
Dr. Pepper	5	
	25	

3. The administration at a high school obtained a random sample of 100 absences. The following table gives the data collected for this sample. Assume the absences occur on the five days with equal frequency; find the expected number and calculate chi-square.

Day	Number absent	Expected number absent
Monday	25	
Tuesday	17	
Wednesday	24	
Thursday	22	
Friday	12	
	100	

114   Chapter 6   Inference

4. The manager of an ice cream parlor claims that customers have no preference between different flavors of ice cream. A random sample of 92 customers showed the following preferences:

Flavor	Frequency observed	Frequency expected
Vanilla	25	
Chocolate	20	
Strawberry	14	
Butter pecan	17	
Mint chocolate	16	
	92	

Find the expected frequency and calculate chi-square.

6.5 Interpreting Chi-Square

In the last section, we calculated a chi-square statistic of 4.75 for the blue jean example. What does this mean? Recall that chi-square is an index used to measure the difference between the observed and expected value. Large differences suggest that each observation is not equally likely to occur. A small value of χ^2 suggests that the observed numbers are very similar to the expected values. For the blue jean example, a small χ^2 would mean that one brand is not really preferred over any other type.

The question that remains to be answered is whether the chi-square value of 4.75 is large enough to conclude that customers do have a preference. In other words, if the customers are choosing a jean type purely at random, will the difference between the observed and expected numbers be large enough to produce by chance a χ^2 value as large as or larger than 4.75? To answer this question, randomly generate a set of χ^2 values, and then compare 4.75 to this random distribution.

A random chi-square for selecting a jean type can be obtained by using the following program:

6.5 Interpreting Chi-Square

PRGM2: COMP
:Disp "SAMPLE"
:Disp "SIZE"
:Input N
:Disp "CELLS"
:Input C
:C → Acol   (on [2nd] [VARS] DIM menu)
:1 → Brow
:0 → [A]
:0 → T
:Lbl 1
:Int (C * Rand + 1) → Y
:[A] (1,Y) + 1 → [A](1,Y)
:T + 1 → T
:If T < N
:Goto 1
:Disp [A]

To execute the program, enter a 32 for the sample size and a 4 for the number of cells. The screen will display an array with four numbers. A sample run produced the following output:

[8 10 6 8].

These numbers represent the observed values that were randomly generated (i.e., assuming no preference between the different types of jeans).

Using these four values, execute Prgm1: CHI2 (program from Section 6.4). Remember to enter 4 for the number of CELLS, 32 for SAMPLE and 8, 10, 6, and 8 after the OBSERVE prompt. The value of χ^2 for this simulation of a random selection of jeans is 1.0. But, many more values of χ^2 are needed to decide if 4.75 is consistently larger than randomly generated χ^2.

Prgm2: COMP will randomly generate observed values and Prgm1: CHI2 will calculate χ^2 on these observed values. Following is a list of 29 chi-square values that were calculated using the two programs:

4.25, 1, .75, 1, .75, 1.75, 4.75, 6.5, 1, 1.25, .75, 6.25, 9.25, .75, 3.75, 4.75, 4.25, 2.25, 3.25, 3.75, 5.5, .25, 5.25, 2.5, 1.25, .25, .75, 2.25, and 4.25.

Figure 6.8 is a histogram of the χ^2 values displayed in the viewing window [-2, 10] by [-2, 10], xscl = 1, yscl = 1.

From this simulation, the probability that a chi-square of 4.75 or greater can occur by chance was 7 out of 29 trials, or about 24%. This probability is represented by the

116   Chapter 6   Inference

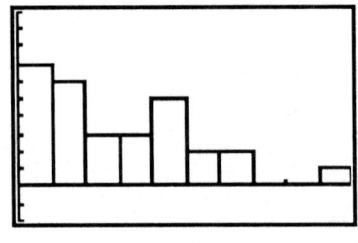 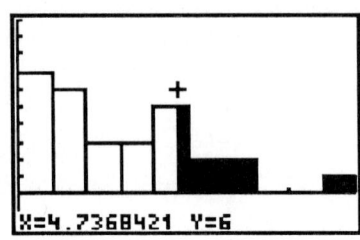

FIGURE 6.8   FIGURE 6.9

shaded region on the histogram in Fig. 6.9. Since a 23% chance is not all that unlikely, this seems to indicate there is no statistical evidence that one jean type was preferred over another. Generally, the probability should be less than 10% to conclude that there is a difference not due to chance. For further information on interpreting chi-square, see *Using Statistics* by Travers *et al.*

Exercises

1. A tetrahedral die was rolled 40 times and the following results were obtained:

Outcome	Frequency	Expected frequency
1	8	
2	8	
3	10	
4	14	
	40	

   a) Complete the expected frequency column in the given table.
   b) Use the CHI2 program from Section 6.4 and calculate chi-square.
   c) Use the COMP program from this section to enter 40 for sample size and 4 for number of cells. List the output.
   d) Use the CHI2 program and the four values generated in part (c) and calculate the value of chi-square.

6.5 Interpreting Chi-Square

e) Repeat parts (c) and (d) to obtain five values of χ^2.

f) Combine your results with six other students. Display the results in a histogram, and calculate the probability that the chi-square value obtained in part (b) would occur by chance.

g) Based on your answer from part (f), do you feel this tetrahedral die is fair? Justify your claim.

2. In Exercise 2, Section 6.4 you calculated a value of χ^2 for the data listed in the following table:

Type of soft drink	Observed	Expected
Coke	5	6.25
Pepsi	6	6.25
7-Up	9	6.25
Dr. Pepper	5	6.25

a) Use the COMP program with 4 for the number of cells and 20 for the sample size. List the output.

b) Use the CHI2 program and the four values from part (a) and calculate the value of chi-square.

c) Repeat parts (a) and (b) until you have obtained five values of χ^2.

d) Combine your results with six other students, display the results in a histogram, and calculate the probability the chi-square value you found would occur by chance.

e) Based on the results from part (d), do you think there is a preference between the different types of soft drinks? Justify your claim.

3. Using the method outlined in problem 2 and referring to problem 3 in Section 6.4, do you think that absences occur with equal frequency? Justify your claim.

4. Referring to problem 4 in Section 6.4, do you think there is a preference between the flavors of ice cream? Justify your claim.

References

Freedman, David, Robert Pisani, and Roger Purves. *Statistics*. New York, NY: W. W. Norton and Company, Inc., 1980.

Gnanadesikan, Mrudulla, Richard Scheaffer, and James Swift. *The Art and Techniques of Simulation*. Palo Alto, CA: Dale Seymour Publications, 1987.

Hoffman, Mark, ed. *The World Almanac and Book of Facts 1988*. New York, NY: Random House, 1988.

Hoffman, Mark, ed. *The World Almanac and Book of Facts 1990*. New York, NY: Random House, 1990.

Landwehr, James, and Ann E. Watkins. *Exploring Data*. Palo Alto, CA: Dale Seymour Publications, 1986.

Landwehr, James, James Swift, and Ann E. Watkins. *Exploring Surveys and Information from Samples*. Palo Alto, CA: Dale Seymour Publications, 1987.

Moore, David, and George McCabe. *Introduction to the Practice of Statistics*. New York: NY: W. H. Freeman, 1989.

Statistical Abstract of the United States 1989. U.S. Department of Commerce. Bureau of the Census. Washington, DC: U.S. Government Printing Office, 1989.

Travers, Kenneth J., William Stout, James Swift, and Joan Sextro. *Using Statistics*. Menlo Park, CA: Addison-Wesley Publishing Company, 1985.

Index

Approximation for $n!$, 76
Average, 7–9

Binomial probability distribution
   discussion of, 76–79
   mean and standard deviation of, 78, 94
   normal distribution used as approximation to, 93–95
Binomial probability formula, 76
Bivariate data
   fitting a line to data and, 34–36
   line $y = x$ and, 30–31
   plots over time and, 27
   scatterplots and, 23–25
Box plot, 99

Chi-square
   calculation of, 111–112
   explanation of, 110–111
   interpretation of, 114–116
Combinations of r items selected from n items, 74
Continuous probability distributions, 86. *See also* Normal probability distribution
Correlation
   explanation of, 39–41
   between two variables, 39
Correlation coefficient
   explanation of, 39
   use of, 43
Curve fitting
   correlation and, 39–41
   nonlinear models and, 49–52
   residuals and, 47
   use of optional, 56–59
   variation and linear data and, 43–45

Data. *See* Bivariate data; Univariate data
Discrete probability distributions, 80–83
Distributions. *See also* Probability distributions
   analysis of, 17, 18
   hypergeometric, 80, 82
   Poisson, 80, 83

ERROR program, 59
ExpReg, 50

Histogram
   creation of new, 3
   probability, 78
   of sample means, 103–104
   use of, 1–3, 63, 67, 82, 116
Hypergeometric distribution, 80, 82

Inference
   chi-square and, 110–112, 114–116
   sampling distribution and, 97–99, 102–105
   testing a claim about a mean and, 105–107
Int function, 66–67
Interquartile range, 12–13

Least squares regression line
   correlation and, 39–41, 43
   curve fitting and, 56
   use of, 34–36
Likely/unlikely samples, 99
Line $y = x$, 30–31
Linear data, 43–45
LnReg, 50

Mean
   for binomial distribution, 78
   explanation of, 7, 8
   finding the, 8
   sampling distribution of sample, 102–105
   standard deviation from, 14–15
   testing a claim about a, 105–107
   variation of, 14–15
Mean squared error, 43
Median
   calling up, 8
   explanation of, 7
   finding the, 7–8

Median (*continued*)
  use of, 8–9
  variation of, 11–13
Mode, 7

n!, 76
Nonlinear models, 49–52
Normal curve
  area under, 89–92
  graphing, 86
Normal probability distributions
  as approximation to binomial distribution, 93–95
  explanation of, 86
  use of, 86–88

Permutations of *r* items selected from *n* items, 73–74
Plots over time, 27
Poisson distribution, 80, 83
Probability
  equal to ½, 61-63
  not equal to ½, 66–68
  and use of simulation, 61. *See also* Simulation
Probability distributions
  and approximation for *n*!, 76
  and area under normal curve, 89–92
  binomial, 76-79
  and combinations of *r* items selected from *n* items, 73–74
  continuous, 86. *See also* Normal probability distribution
  discrete, 80–83
  normal, 86-88
  normal used as approximation to binomial, 93–94
  permutations of *r* items selected from *n* items and, 73–74
Probability histogram, 78
Program Simpson, 89–91
PwrReg, 50

Quartiles, 12–13

Rand function, 61
Residuals
  explanation of, 43, 47
  graphing, 47, 50

Sample mean, 102–107
Sampling distribution
  explanation of, 97–99
  of sample means, 102–105
Scatterplots
  explanation and use of, 23–25
  and fitting a line to data, 34–36
  having two variables based on same scale, 30–31
Seed, 61
Sigma notation, 8
Simpson's method, 89–91
Simulation
  explanation of, 61
  and probability equal to ½, 61–63
  and probability not equal to ½, 66–68
  and unknown number of events, 70–71
  used in sampling distribution, 97–99
Standard deviation
  for binomial distribution, 78
  calculation of, 14–15
  conclusions regarding, 17
  of distribution of sample means, 105
  explanation of, 14
  population, 105
Standard error
  explanation of, 43
  finding the, 43–45
  residuals and, 47
Standard normal distribution, 87–88
Stirling's approximation for *n*!, 76

TRANSFORMATION program, 56–59

Univariate data
  averages and, 7–9
  decision making regarding, 17–18
  making sense of, 1–3
  variation and mean and, 14–15
  variation and median and, 11–13

Variation
  and linear data, 43–45
  of mean, 14–15
  of median, 11–13

01 MATH, 94